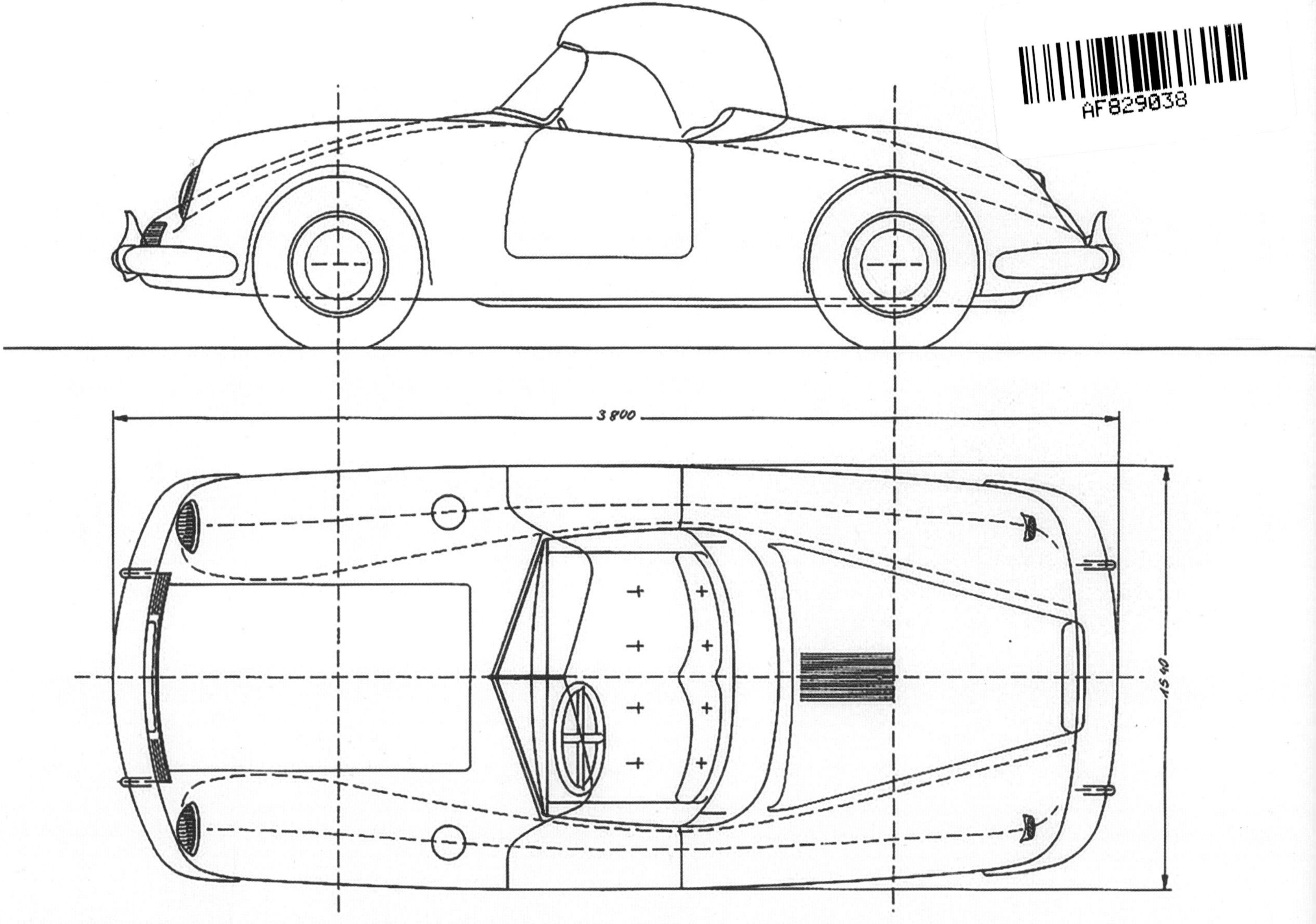
3 800
1540

1931 — PORSCHE CHRONICLE — 2023

EDITION PORSCHE MUSEUM

About this book

Driven by Dreams: Driven by a dream, the history of Porsche began in Gmünd, Austria. That's where the first sports car with the name was built, back in 1948, as Ferry Porsche and his team sought to make his concept for a sports car a reality. At the heart of this process was the pursuit of that vision, regardless of external circumstances, to make what seemed impossible come true.

On 8 June 1948, the first Porsche obtained its general operating permit. Since then, this date has marked the birth of a brand that's much more than just a name. Porsche has become a byword for dynamic driving and a sporting approach – for automotive innovation, design and functionality.

But long before that, Ferdinand Porsche had already laid the foundation for today's company, Dr. Ing. h.c. F. Porsche AG. In 1931, he set up his own engineering office in Stuttgart, helping to make history in the automotive industry with the design of the "Volkswagen", or "people's car".

Over the past decades, the company has experienced many highs and lows and made history in the car-making industry, as well as business as a whole. Thanks to efficient production methods, qualified employees, a clear brand strategy and innovative models, what once was a small sports car workshop in Stuttgart-Zuffenhausen has become a world-leading manufacturer of premium sports cars in technology and style – and one of the industry's most profitable businesses to boot.

The Porsche Chronicle presents a timeline of highlights from the brand's fascinating history in words and images. From its beginnings in 1931 to the first Porsche sports car in 1948, and the next 75 years that brought us to the 2023 anniversary year – it's ideal for those looking for the low-down on the world-famous brand with Swabian roots.

Foreword

It all began with a dream. Ferry Porsche dreamt of an extraordinary sports car. And since no such car existed, he and his team built it themselves. On 8 June 1948, his 356 "No. 1" Roadster obtained its general operating permit. This day marked the birth of the sports car brand Porsche – and the beginning of a unique success story that is constantly adding new chapters.

Its historic roots go back even further, however. After all, Ferry's father himself had dreams as well – and made automotive history by making them come true. As a young man, Professor Ferdinand Porsche built electric and hybrid-powered vehicles that attracted attention. That was around 120 years ago. Back then, he dreamt of his own business – and, together with Anton Piëch and Adolf Rosenberger, he made this dream come true.

The Porsche engineering office created the designs and prototypes for cars like the Volkswagen "Beetle". Tractors and aircraft engines were also designed here. Racing cars were a special focus. Putting innovative designs to the test in sporting competition was part of Ferdinand Porsche's philosophy from day one. Ferry Porsche took inspiration – and high standards for his own work – from this family background. This was where the seeds for Porsche and its subsequent development as a sports car manufacturer were sown.

In 2023, Porsche is celebrating its anniversary. Our brand can now look back on a proud tradition spanning 75 years. A fascinating history with many highs and a few lows, too. With countless motorsporting triumphs and all kinds of technical innovations. With unique design icons, shaped by a distinctive company culture. A history of passion and pioneering spirit, a love of sport and a sense of family. Driven by the desire to make our customers' dreams come true.

Today, Porsche is a strong global brand. It has long since transformed from a small sports car workshop to a world-leading manufacturer of exclusive sports cars. A brand that epitomises modern luxury – and combines performance with innovation and sustainability. A brand with a pioneering ambition to help shape the future

of mobility. The rise of Porsche can only be understood by looking to the past: over the years, Porsche has evolved and transformed, time and time again. That's the only way Porsche could remain Porsche.

This compendium traces the history of those innovations – with a timeline that sums up the key milestones, making for an entertaining read or a quick reference guide. But the Porsche Chronicle doesn't just look back – it points the way to the future, too. It shows what sets the success and star power of Porsche apart – yesterday, today and tomorrow.

OLIVER BLUME

Porsche: company and brand

Company founders

Ferdinand Porsche, Adolf Rosenberger, Anton Piëch

25/04/1931

Porsche engineering office

The company "Dr. Ing. h.c. F. Porsche Gesellschaft mit beschränkter Haftung, Konstruktionen und Beratung für Motoren und Fahrzeugbau" was entered into the City of Stuttgart's Commercial Register.

1937

Transition to a limited partnership (KG)

Transition of Dr. Ing. h.c. F. Porsche GmbH to Dr. Ing. h.c. F. Porsche KG, Stuttgart in December 1937.

1944

Relocation

Relocation to Gmünd, Carinthia in Austria due to the war.

Founder of the Porsche brand

Ferdinand Anton Ernst "Ferry" Porsche

08/06/1948

Birth of the brand

The Porsche 356 "No. 1" Roadster: the first car to bear the name Porsche was awarded its general operating permit.

1949

Return

Return to Stuttgart in November 1949 and start of series production of the 356.

1952

Origin of a trademark

Porsche received an identifying trademark.

1972

Transition to a limited company (AG)

Transition of Dr. Ing. h.c. F. Porsche KG, Stuttgart to Dr. Ing. h.c. F. Porsche AG.

Content

Ferdinand Porsche (left) and son Ferdinand Anton Ernst "Ferry" in the early 1930s.

1931 — FOUNDING OF THE COMPANY — 1939

“I’d be better off setting up my own office and working for everyone.”

Ferdinand Porsche

1930s 1950 1960 1970 1980 1990 2000 2010 2020

1931

The company "Dr. Ing. h.c. F. Porsche Gesellschaft mit beschränkter Haftung, Konstruktion und Beratung für Motoren- und Fahrzeugbau" was entered into the Stuttgart Commercial Register on 25 April at the height of the global economic crisis. It was named after its founder Dr. Ing. h.c. Ferdinand Porsche. The engineering office based at Kronenstrasse 24 in Stuttgart had been set up with 12 employees in December 1930. Alongside Ferdinand Porsche, who invested 24,000 Reichsmarks in the limited company's share capital, his stepson Anton Piëch and Adolf Rosenberger each took a stake of 3,000 Reichsmarks as managing partners. The company boasted a wide range of designs from the outset.

The engineering office was awarded its first contract by the Chemnitz-based automobile manufacturer Wanderer, which requested the development of a six-cylinder mid-range saloon and a new in-line eight-cylinder engine. This was followed by a swing axle for Horch in Zwickau and an air-cooled five-cylinder radial engine commissioned by Phänomen in Zittau to power trucks.

Anton Piëch (left) and Adolf Rosenberger.

The first design commission for the Porsche engineering office came from Wanderer (Porsche Type 7).

Illustration of the "System Porsche" torsion bar suspension designed under the designation Type 17.

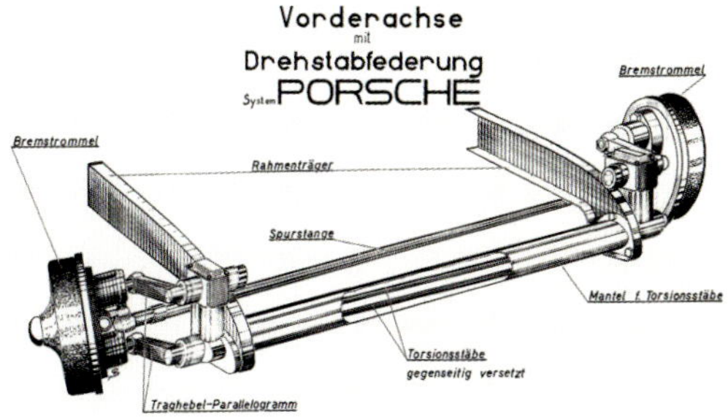

A technical highlight of the first year's work and milestone of technical history was the torsion bar suspension. A patent was applied for this component on 10 August. It is still used in international automotive manufacture today.

From September, the Porsche engineering office began work on a small car on behalf of Zündapp GmbH. The Type 12, as it was known, was set to influence the later Volkswagen with its rear engine, central tube frame and a gearbox in front of the rear axle.

Illustration of the Type 12 small car commissioned by Zündapp GmbH, 4 April 1932.

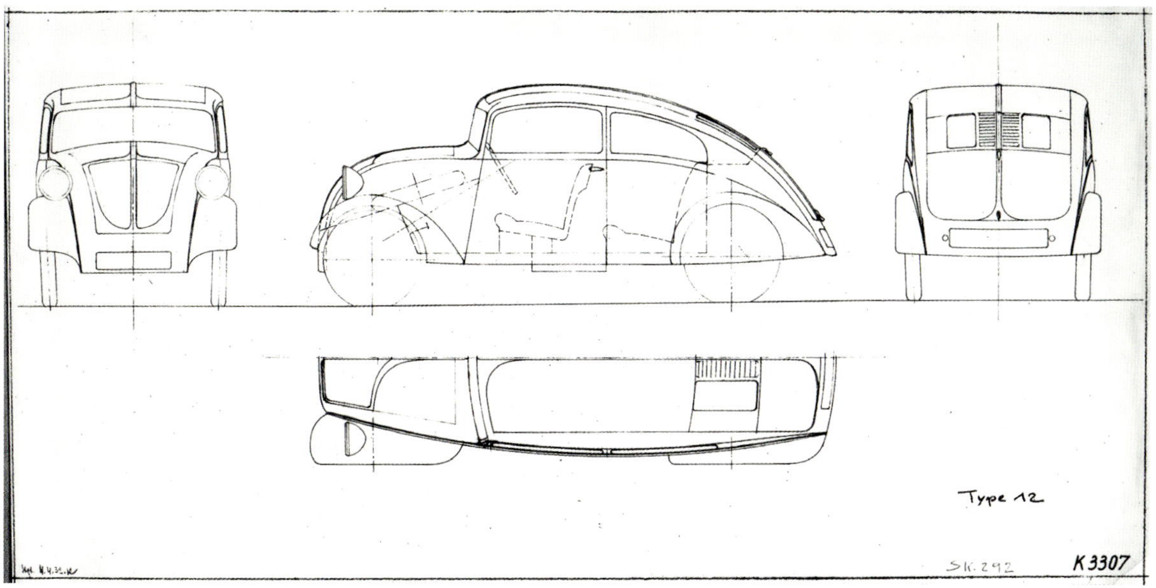

1932

The general economic depression was not without consequences for the Porsche engineering office – and 1932 saw the first test of the fledgling company's resilience. Although work on Type 12 was completed in spring, Zündapp opted for economic reasons not to put this promising small car model into serial production, as they wished to focus solely on motorbike manufacturing.

Ferry Porsche with the Porsche Type 12 Cabriolet, developed for Zündapp.

As a world-renowned automotive design engineer, Ferdinand Porsche received a Russian delegation who offered him the chance to help develop the Soviet automotive industry as the General Director for Design and Development. After spending several weeks in Russia and visiting numerous industrial facilities, the 57-year-old turned down the lucrative role – he felt the linguistic and cultural obstacles were too great.

Ferdinand Porsche.

When the international racing authorities announced the new 750 kilogrammes Formula, Ferdinand Porsche saw the opportunity to put his many years as a racing car design engineer to the test once again. A subsidiary was founded to reduce the financial risks of a capital-intensive racing car project. It also traded at Kronenstrasse 24. The Hochleistungs-Fahrzeug-Bau GmbH company was entered in the Commercial Register on 19 November and started out by designing a new high-performance 16-cylinder engine for a future Grand Prix racing car.

1933

After lengthy negotiations, agreement was reached on 17 March with the Saxon car manufacturer Auto Union on the construction of a 750 kilogrammes Formula racing car. Supervised by chief engineer Karl Rabe, the small team of Porsche technicians made such strides in developing the Type 22 mid-engined car within a few months that work was able to begin on building the first P-Rennwagen racing car (P for Porsche) at Auto Union in Chemnitz that spring. Alongside its prestige, the Type 22 design was also so financially lucrative for Porsche that it was possible to liquidate the Hochleistungs-Fahrzeug-Bau GmbH company, previously established as a safety measure.

Alongside the Type 22 Auto Union racing car, a small car ordered by NSU became the year's most crucial development order. The design works began in August and involved a rear-mounted air-cooled four-cylinder boxer engine and Porsche torsion bar suspension for the Type 32. Following approval of the designs by NSU director Fritz von Falkenhayn on 10 November, two prototypes were produced by Heilbronn-based chassis maker Drauz in a mixed construction with a wooden chassis and artificial leather casing. Reutter, a bodyshop factory based on Stuttgart's Augustenstrasse, produced an all-steel chassis for a third prototype.

Test-driving the Auto Union racing car Type 22, November 1933.

Volkswagen Type 32 outside the Porsche engineering office at Kronenstrasse 24.

1934

Due to the difficult economic situation, the idea of a low-cost small car or "Volkswagen" (people's car) had already been examined time and again by various automotive design engineers since the late 1920s. On 17 January, at the Reich Ministry of Transport in Berlin, Ferdinand Porsche summarised his own ideas on the matter in an "Exposé on the construction of a German people's car". The Type 60 outlined here was already the eighth small car developed under his direction. Furthermore, the commissioned work performed in the Porsche engineering office for Zündapp and NSU already showed technical and stylistic elements of what would later become the "Volkswagen" people's car.

Ferdinand Porsche at the drawing board in the Stuttgart engineering office.

On 22 June, Dr. Ing. h.c. F. Porsche GmbH was officially awarded the contract to design and build a German people's car, or "Volkswagen", by the Reich Association of the German Automotive Industry (RDA). Initially, the manufacture of a prototype was agreed. The number of test cars was contractually increased to three vehicles on 7 December 1934.

Alongside working on the Volkswagen project, the engineering office also supervised the Auto Union P-Rennwagen developed as the Porsche Type 22 during the 1934 racing season. Test drives with Ferdinand Porsche in attendance took place at the Berlin AVUS test track as early as January. During its first year in use, the Auto Union P-Wagen racing car set three world records and won the Grand Prix in Germany, Switzerland and Czechoslovakia, as well as four international hill climbs.

Racing driver Hans Stuck at the Brno Grand Prix in an Auto Union Type A (Porsche Type 22).

1935

By 1 January, four years after the Porsche engineering office was founded, the workforce had grown to 35 employees. The office invested all its energy in the Volkswagen project, facing resistance from Germany's automotive industry, which feared competition with its existing vehicle models.

The Volkswagen prototype Porsche Type 60 (V1) during test drives in winter 1935/36.

Ferry Porsche behind the wheel of the second "Volkswagen" prototype (V2).

Two Volkswagen Type 60 prototypes were created in the garage of the Porsche estate at Feuerbacher Weg in Stuttgart. The first test car, known as V1, was presented to a committee of the Reich Association of the German Automotive Industry (RDA) on 3 July 1935. The second test car, the V2, made its maiden journey on 22 December and was presented to Adolf Hitler together with the V1 prototype in Munich on 29 December.

Alongside the car's development, from December onwards, Porsche GmbH began work on other areas: in addition to the design of a 1,000 PS aircraft engine (Type 55) for Munich-based Süd-Bremse AG, the German Aerospace Test Institute (DVL) ordered a 32- and 16-cylinder aircraft engine with the internal designation Type 70 and Type 72 respectively.

Other projects included swing axle design for the British racing team English Racing Automobiles (ERA) and a motorcycle study for the Nuremberg-based motorbike manufacturer Zündapp.

1936

While work began on building three additional Volkswagen prototypes known as V3 in February, there was increasing resistance on the part of the Reich Association of the German Automotive Industry (RDA). Scheduling delays and various technical demands for the test cars led to heavy criticism from RDA president Robert Allmers. Contrary to the initial idea of having the Volkswagen built as a joint project between German automotive manufacturers, Adolf Hitler decided on 4 July to build an independent Volkswagen plant which was to be planned by Ferdinand Porsche.

To find out about American production methods, Dr Porsche and his nephew and private secretary Ghislaine Kaes headed to the US for a six-week study visit on 1 October. In Detroit, he visited the modern production facilities of the Ford Motor Company and General Motors Corporation. Back in Germany, Porsche received the Wilhelm Exner medal awarded by the Austrian Trade Association (ÖGV) for "outstanding contributions to automotive manufacture" on 18 December.

Ferdinand Porsche on board the Queen Mary in New York Harbour.

By mid December, the V3 Volkswagen prototypes had covered 50,000 kilometres each in their extensive endurance test drives without any fundamental issues emerging. Technical observers from the RDA took their seats in the test cars alongside drivers from Porsche GmbH.

Porsche Type 60 (V3) at the courtyard exit of the Porsche Villa in Stuttgart.

1937

Alongside the development of technical vehicle components, on 11 March, Daimler-Benz AG commissioned the design of the Type 80 "Rekordwagen", which was intended to break the land speed record with a special aircraft engine with 3,500 PS. At the same time, Daimler-Benz was working on a test series of 30 Volkswagen Type W30 prototypes, commissioned by the RDA. A year of road testing by an SS roadworthiness team began on 15 May with technical supervision from Ferdinand Porsche's son Ferry and covered over 2.4 million test kilometres.

The Mercedes-Benz high-speed record car T 80, built by Porsche.

On 28 May, in Berlin, with backing from the German Labour Front (DAF), the Gesellschaft zur Vorbereitung des Deutschen Volkswagens mbH (Gezuvor) was established. As one of three managing directors, Ferdinand Porsche was now officially responsible for the technology and planning of the future Volkswagen plant.

Accompanied by son Ferry and nephew Ghislaine Kaes, Porsche headed to the US for a second study visit on 26 June, together with two other Gezuvor managing directors, Bodo Lafferentz and Jakob Werlin. The aim of the trip was to purchase special machinery and recruit production specialists for the Volkswagen plant.

Ferdinand Porsche and son Ferry on a study visit to the US.

The design of the small tractor Type 110, begun on 24 November, featuring an air-cooled 12 PS two-cylinder engine, later became the basis of the subsequent "Volkstraktor" and the Porsche diesel tractors produced after the Second World War.

Alongside the Volkswagen, 1937 saw the creation of the Type 110, a forerunner of the later "Volkstraktor".

At the end of 1937, Dr. Ing. h.c. F. Porsche GmbH became a limited partnership (Kommanditgesellschaft, or KG). Joining Managing Director Dr Ferdinand Porsche were additional partners Anton Piëch, Baron Hans Veyder von Malberg, Ferry Porsche and Louise Piëch.

1938

For Dr. Ing. h.c. F. Porsche KG, 1938 began with the design contract for a double compressor for the 12-cylinder engine of the Mercedes-Benz three-litre Formula racing car W 154 and the Type 90, a design for a racing car which would never be realised.

Technical drawing of the prototype racing car Type 90, 12 April 1937.

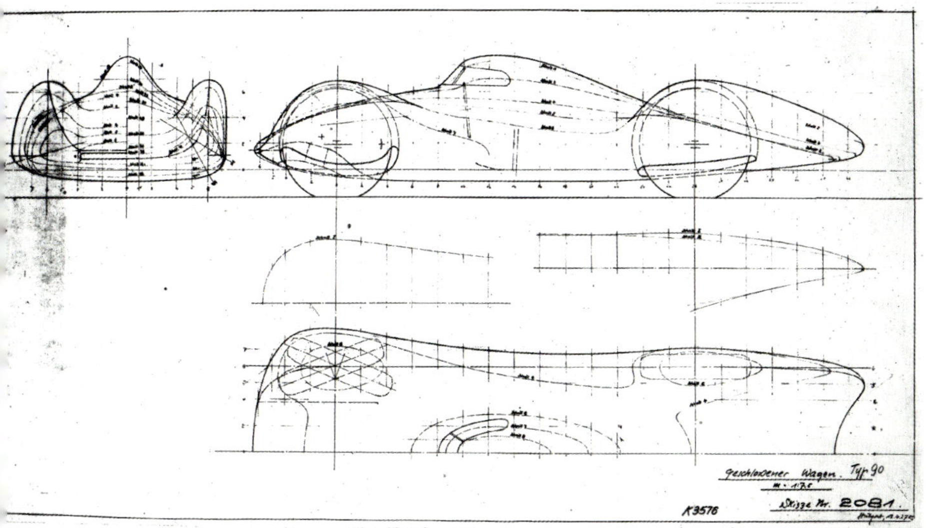

In spring, a second series of thirty Volkswagen Type VW 38 test cars was created and supplied with structures from the Stuttgart-based chassis works Reutter & Co. Each week saw the production of two chassis which had reached their final shape with the doors hinged to the A-pillar and what was known as a "pretzel window".

Ferdinand Porsche (left) and Karl Rabe discussing a drawing of the VW 38.

From June onwards, comparative test drives with other series vehicles were held for the Volkswagen, now renamed the KdF car (strength through joy). In the first week of July, Ferry Porsche took part in an Alpine test drive with a V 303 Volkswagen saloon and a cabriolet. Towards the end of the month, a second test drive took place with the VW 38, now fully developed.

The Porsche engineering office moved out of Kronenstrasse 24 on 26 June and into a dedicated building located in Stuttgart-Zuffenhausen. Ferry Porsche had purchased the plot of land at Spitalwaldstrasse 2 on behalf of the company in May 1937, thereby determining the subsequent location of Porsche Werk 1.

The VW 38 prototypes of the Porsche Type 60 from 1938 already shared the design of the subsequent Volkswagen.

In September, the team at Porsche came up with an innovative 1.5-litre mid-engined sports car – the Type 114 project. Technicians opted for a water-cooled, ten-cylinder V-engine for the racing car Type 116 derived from this. However, the outbreak of war saw all plans to implement these projects come to nothing. A happier fate was in store for the Type 64, intended for the endurance rally from Berlin to Rome, whose design had begun in September. It was the ancestor of all subsequent Porsche sports cars.

Porsche Werk 1 at Spitalwaldstrasse 2 during construction, 1938.

A Porsche Type 114 model from 1938.

1939

In spring, Dr. Ing. h.c. F. Porsche KG developed and built three racing coupes known as the Type 64 and Type 60K10 for the long-distance Berlin to Rome rally planned for September 1939. The design and aerodynamics were far ahead of their time. With a streamlined aluminium chassis and a modified Volkswagen boxer engine with 50 PS, the Berlin-Rome racing car reached top speeds of 145 km/h. When the outbreak of war prevented the race from taking place, the Porsche engineering office used the finished racing cars as rapid travel vehicles reaching average speeds of over 130 km/h.

In 1938/39, the Type 64, developed for long-distance races, was created on the basis of the Volkswagen.

134

Based on the KdF car presented at the Berlin Motor Show in February, additional vehicle types were prepared for military use over the course of year. Alongside the Type 81 "Kastenwagen" box van, Porsche KG developed the Type 62 off-road vehicle, the Type 82, which became known as the "Kübelwagen", and the all-wheel-drive Type 87.

Alongside the vehicle drive, the air-cooled VW engine also served as a stationary engine that was designed in 1939 for various uses with magneto and battery ignition systems.

In early December, the Porsche engineering office received its first order to develop a mid-weight combat tank from the German Army's Armaments Office (Heereswaffenamt). Work on designing the Type 100 "Leopard", whose hybrid petrol and electric drive featured two air-cooled V10 diesel engines, was finished in April 1941. However, as a more heavy-duty tank was required in late May 1941, all further work on the Type 100 came to an early end.

Test-driving the mid-weight Type 100 "Leopard" combat tank.

The all-wheel-drive Type 87 off-road vehicle, designed by Porsche, on a winter test drive.

Ferdinand Porsche with his grandsons Ferdinand Piëch (right) and Ferdinand Alexander Porsche.

1940 — FOUNDING OF THE BRAND — 1949

"In the beginning, I looked around and could not find quite the car I dreamed of. So I decided to build it myself."

Ferry Porsche

1930 1940s 1960 1970 1980 1990 2000 2010 2020

Testing the amphibious properties of the Type 128 in the Porsche KG fire extinguishing basin.

1940

In June, Porsche KG received a request from the German Army's Armaments Office to develop an amphibious off-road vehicle based on the four-wheel-drive Type 87 "Kübelwagen" all-terrain vehicle. The first Type 128 "VW-Schwimmwagen" test vehicle was initially trialled in the fire extinguishing basin at the Porsche site in Zuffenhausen. In early November, three more Type 128 "Schwimmwagen" vehicles were tested on an extensive test run.

Throughout the year, Dr. Ing. h.c. F. Porsche KG developed an array of technical detailed solutions for the KdF car, including a fully synchronised gearbox and a hybrid petrol-electric heating system.

The entrance to the Porsche plant in Stuttgart-Zuffenhausen, with VW prototype W30.

In August, the Porsche engineering office began developing a series of wind power facilities, commissioned by the Berlin Wind Power Research Association (Berliner Forschungsgesellschaft für Windkraft). Three wind turbines with power classes of 130 watts (Type 135), 736 watts (Type 136) and 4,500 watts (Type 137) were built and erected for testing purposes in Zuffenhausen, Schwäbisch Gmünd and Hohenheim. The three-blade rotor of the biggest appliance, Type 137, had a diameter of 9.2 metres and was fitted on a 17-metre-tall steel frame.

Testing Type 137 wind turbines.

Porsche engineers developed a motorcycle rear wheel suspension and a wood gasification plant for Austria's Steyr-Werke and inspected an eight-cylinder Steyr passenger vehicle and an all-wheel-drive army truck.

The Reich Ministry for Science and Education gave Ferdinand Porsche the title of honorary Professor in September 1940 to mark his 65th birthday.

1941

The Daimler-Benz AG board recommended that the Armaments Ministry hire the Porsche engineering office as consultants. The Reich Minister for Armaments and Ammunition, Fritz Todt, then made Professor Ferdinand Porsche head of the Panzerkommission (Tanks Commission) in June.

The switch to a wartime economy saw the deployment of coerced workers at the main Zuffenhausen plant between 1941 and 1944. During this period, over 400 foreign and forced workers were deployed. Most were skilled workers who worked in turning, milling or mechanics. They were accommodated in simple wooden barracks on and around the factory site. Forced labourers were also repeatedly requested from internment camps in the area around Zuffenhausen. As a small business with just a few hundred employees, Porsche KG had comparatively more leeway in its treatment of forced labourers than larger companies.

With the Type 101 "Tiger", Porsche KG took part in an invitation to tender on the part of the German Army's Armaments Office for the design of a tank in the weight class over 50 tonnes. However, there wasn't enough time to get the brand-new, air-cooled ten-cylinder diesel engine ready for production, so the competing company Henschel won the tender.

The Type 101 on test drives in 1941.

The Type 112 "Volkstraktor" was an evolution of the Type 111 tractor with a larger engine and appropriately adapted gearbox. The air-cooled two-cylinder V-engine had a capacity of 1,700 cc and power of 15 PS.

A Porsche "Volkstraktor" Type 112 from 1941.

The testing and development of Type 128 "Schwimmwagen" and Type 138 all-terrain vehicles led to the design of the smaller test vehicle Type 166 "Kradschützenwagen" (light personnel carrier) with a shortened wheelbase. Weighing only 900 kilos thanks to its compact design, the Type 166 was intended as an alternative to motorcycle and sidecar combinations.

The Type 166 emerged as a shortened version of the Type 128 "VW-Schwimmwagen".

1942

In the absence of military traction machines with high ground clearance, Porsche KG received the contract to develop the Type 175 "Ostrad-Schlepper" in January. With its outsize wheels, Type 175 was similar to the Austro-Daimler mortar truck designed by Ferdinand Porsche in the First World War.

Ferdinand Porsche after a test drive with the Porsche Type 175 "Ostrad-Schlepper".

In March, Ferdinand Porsche was awarded the contract to design a heavy-duty Type 205 "Maus" tank, intended for use as a mobile bunker system with armour of 240 millimetres. Once drafting was complete, the tank with its hybrid petrol-electric drive weighed a mighty 189 tonnes overall. In total, only two prototypes of the "Maus" were built. They were tested in 1944 and 1945.

Based on the Type 101 "Tiger" tank, an anti-tank vehicle equipped with two water-cooled Maybach diesel engines was developed with the internal designation Type 131 "Ferdinand".

Type 131, "Ferdinand".

Work began on an air-cooled 16-cylinder diesel engine with direct fuel injection. Designated Type 220, it was intended to offer up to 800 PS with a 36.8-litre capacity and exhaust gas turbocharging. However, only two Type 220 test engines were produced by the end of the war.

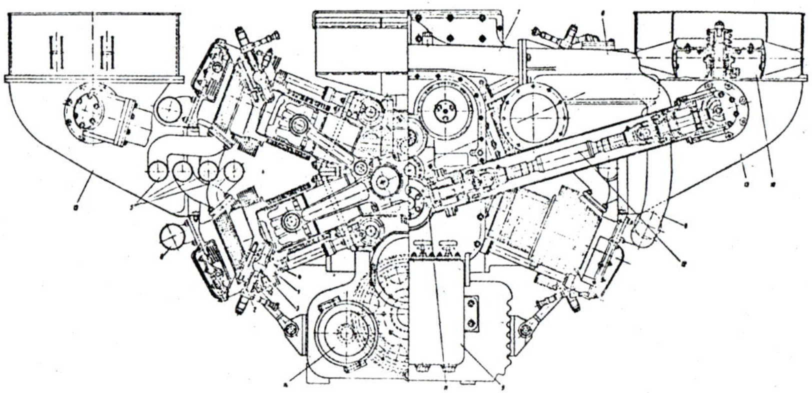

Illustrated cross section of the Porsche Type 220 tank engine.

1943

Air-raid alerts and aerial attacks made work on Porsche KG's development contracts tougher in 1943. The Type 205 "Maus" anti-tank vehicles were top priority. In this context, Porsche worked closely with Daimler-Benz AG, who contributed to the 1,080 PS, water-cooled in-line 12-cylinder engine Type DB 509. A 1:1 scale wooden model of the "Maus" was presented to Hitler in May. The first prototype was completed in late December and underwent an unofficial test drive. Official testing of the Type 205 "Maus" tank took place from mid January to late October 1944 on the site of the Böblingen tank barracks.

The Type 205 "Maus" was designed as a "portable bunker", but was never used.

For the Type 82 "Kübelwagen", a snow chain mechanism was developed as Type 155, suitable for speedy installation. However, serial production of the snow chain did not take place due to the wartime scarcity of raw materials and resources.

Test-driving the Porsche Type 155 in Zell am See on 7 March 1944.

Test drives with the Type 157 rail vehicle took place in August on the sidings at Stuttgart Hauptbahnhof main railway station. The "Kübelwagen" Type 82 and Type 87 formed the basis of this vehicle. To ensure the test cars could also run on railway tracks, rail guide wheels were fitted inside the wheels.

Following differences with Armaments Minister Albert Speer over fundamental technical issues, Ferdinand Porsche was removed from his position leading the Tanks Commission in October.

Ferdinand Porsche.

1944

More than ever, the lives of the Porsche employees were impacted by the heavy air raids on Stuttgart. At the insistence of the Wehrmacht's Armaments Office, Porsche KG relocated the engineering office from Stuttgart to Gmünd in Carinthia in the autumn. Temporary workshops were set up on the grounds of the company W. Meineke Holzgroßindustrie Berlin-Gmünd under the auspices of chief engineer Karl Rabe. The materials warehouse was located at a flying school in nearby Zell am See. Divided up into numerous work sites and accommodations, the employees gave the new Porsche works the moniker "United Hut Works".

Porsche KG moved into the business premises of a former sawmill in Gmünd, Austria.

One of the first projects achieved in Gmünd was the development of a gas turbine (Type 305) with shaft power of 1,000 PS. However, due to the war, the turbine, which was designed to drive a tank, did not pass the design stage.

In addition, the Porsche engineers also worked on a gas generator system (Type 230) and the development of a new two-stroke diesel engine (Type 309) for the Volkswagen. Designated Type 287, the chassis of the "Kommandeurwagen" was designed in Zuffenhausen based on the all-wheel drive Type 87.

The all-wheel-drive Porsche Type 287 "Kommandeurwagen".

After initially suffering only slight damage from bombings in February and April, the Porsche works on Spitalwaldstrasse fell victim to a major air raid in the night of 19 to 20 October. As well as the Volkswagen spare parts depot, the technical drawing and files archive was hit. It had been moved into the cellar of Building No. 2 from an attic shortly before.

1945

After the war ended, the Dr. Ing. h.c. F. Porsche KG works in Zuffenhausen were initially used by the French military as a transitional camp for Russian prisoners of war requiring repatriation. In August, an American unit took over the works, now repurposed as a repair workshop for trucks.

German employees of the US Army outside Werk 1 in Zuffenhausen.

On August 8, the Porsche works in Gmünd were given provisional authorisation to restart work by the Allied military government in Klagenfurt. The Porsche workforce, amounting to around 140 employees managed by chief designer Karl Rabe, was granted permission to create "designs for motorised tractors, gas generators and other civil facilities" and repair "motor vehicles and agricultural machines". The spectrum of design works ranged from cable machines and water turbines to building fittings and ski bindings.

In early August, Professor Ferdinand Porsche was detained by order of the Allied military government in Gmünd and interned in a special detention camp in Kransberg Castle, Bad Nauheim. He was released in mid September due to exonerating testimony from witnesses.

Ferdinand Porsche a few months before his arrest in 1945.

In mid November, Professor Ferdinand Porsche accepted the invitation of a French commission to go to Baden-Baden and discuss continuation of the Volkswagen project in France.

It was intended that this would be produced by Renault. Renault's competitor Peugeot sought to prevent this and accused Porsche of involvement in war crimes due to Volkswagen factory oversight of Peugeot factories. This accusation saw France's military justice system take Porsche into custody in December 1945. From then on, his son Ferry Porsche controlled the destiny of the business.

1946

Professor Ferdinand Porsche was taken to Paris in May, where, as a prisoner, he initially advised French engineers on the development of the Renault 4 CV.

Switzerland's Rupprecht von Senger contacted the Gmünd Porsche engineering office in June with a request for a four-seater passenger car that was then designed with permission from the British military authorities under Porsche type number 352. A miniature model was produced whose chassis shape, designed by Erwin Komenda, was already strikingly similar to the subsequent Porsche 356.

Since 1931, Designer Erwin Komenda was one of the most important employees of the Porsche engineering office.

In August, the Porsche company took part in the Carinthian Trade and Industry Exhibition. For the first time ever, the product range produced in Gmünd – including water turbines, cable winches, ski lifts, mowing rods and various tractors based on the "Volkstraktor" – was presented under the company's own name.

The Porsche KG booth at the Trade and Industry Exhibition in Klagenfurt on 8–22 August 1946.

In a deal brokered by Karl Abarth and Rodolfo Hruska, a contract was concluded with the Turin-based industrialist Piero Dusio for extensive development orders for the company Cisitalia on 15 December. In addition to a small tractor and a water turbine, Gmünd's Porsche KG began designing the Type 360 Cisitalia Grand Prix racing car and the two-seater Type 370 mid-engined sports car.

A model of the Type 370 for the Italian company Cisitalia.

1947

The design engineers at Porsche KG in Gmünd worked hard on the Type 360 Grand Prix racing car. The vehicle developed for Cisitalia was way ahead of its time – it was to be powered by a 1.5-litre 12-cylinder engine with supercharging, while power transmission was seen to by an engageable four-wheel drive. A two-litre eight-cylinder boxer engine was planned for the Type 370 touring sports car, which was to compete in endurance races such as the Mille Miglia. However, owing to the Italian customer's lack of funds, the Cisitalia racing car never made it beyond the testing stage.

The all-wheel-drive Type 360 was powered by a supercharged twelve-cylinder engine.

To avoid the threat of expropriation of German company property, the Austrian company Porsche Konstruktion GmbH with headquarters in Gmünd was established on 1 April and officially entered in the Commercial Register on 24 April. Managed by

Ferdinand Porsche (left) and Karl Rabe.

Porsche's daughter Louise Piëch and chief designer Karl Rabe, the new company took over the workforce and machinery of the Austrian divisions of Dr. Ing. h.c. F. Porsche KG.

In May 1947, Ferdinand Porsche's protection by the French Ministry of Production came to an end. Porsche was forced to spend several months in military prison in Dijon. As the charges turned out to be unfounded, he was released on bail in July 1947 and officially acquitted in May 1948.

On 17 July, work began on designing the Type 356 "VW sports car". The first draft designs of a body model were drawn up, based on the Type 64 Berlin-Rome racing car from 1938.

Illustration of the 356 "No.1" Roadster, July 1947.

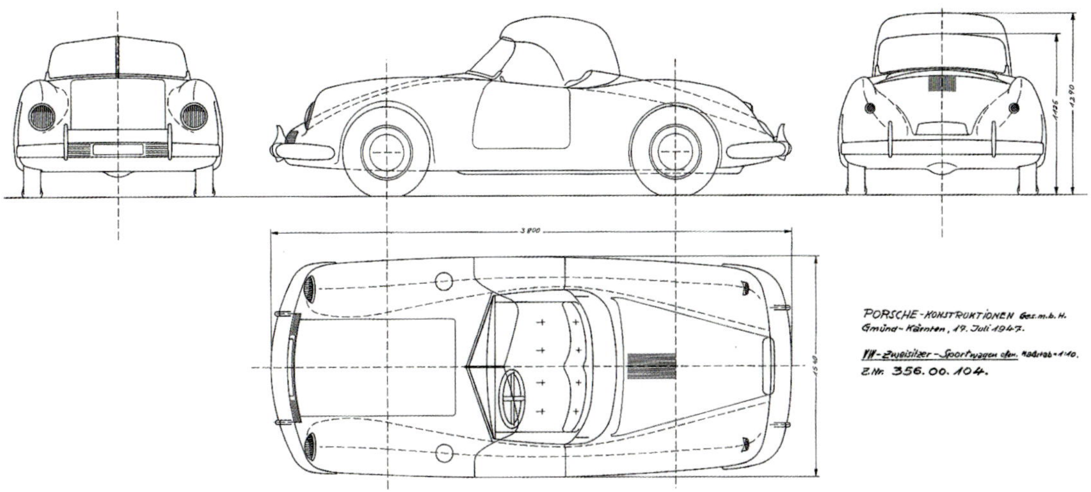

1948

The completed designs for the 356 were realised in spring. The first sports car under the Porsche name made its appearance. On 8 June, the 356 "No.1" Roadster with chassis number 356-001 received its general operating permit from the Carinthian state government. A mid-engined sports car kitted out with a Volkswagen engine boosted to 26 kW (35 PS), weighing in at just 585 kilogrammes, the one-off production reached a maximum speed of 135 km/h.

Ferdinand Porsche (right), son Ferry (centre) and Erwin Komenda (left) with the Porsche 356 "No.1" Roadster.

From the outset, sports car manufacturer Porsche used motorsport as a platform for testing new car types. With Herbert Kaes at the wheel, the Porsche 356 "No. 1" Roadster performed a quick demonstration lap on 11 July at the Innsbruck city race.

Herbert Kaes piloted the 356 "No. 1" Roadster in Innsbruck.

In tandem with the "No.1" development, a coupe version with a rear engine began to take shape. The first Type 356/2 car received type certification from the state governor in Klagenfurt on 7 August. While the Roadster remained a one-off production, as a coupe and cabriolet, the 365/2 formed the basis of the subsequent success of the new sports car brand Porsche.

The manual production of the Porsche 356/2 continued into the second half of the year. Swiss entrepreneur Rupprecht von Senger supplied the necessary capital and missing spare parts and metal. He also snapped up the first five Porsche sports cars

Production of the first Porsche 356 in the main assembly building in Gmünd, Carinthia, circa 1948.

for export to Switzerland. Together with hotelier and car dealer Bernhard Blank, he laid the foundations for subsequent collaboration between Porsche and Switzerland's AMAG Automobil- und Motoren AG.

In September, Ferry Porsche negotiated a forward-looking contract with Heinrich Nordhoff, the managing director and later general director of Volkswagenwerk GmbH, in Bad Reichenhall. The Volkswagenwerk plant had previously been taken over by the British military government in June 1945 and began car production under the management of Major Ivan Hirst. In addition to a future licence fee of five Deutschmarks per completed VW Beetle, Volkswagen now undertook to supply components for Porsche sports cars and also permitted the use of the VW sales organisation and service network. In return, Porsche engineers would be on hand to advise the development department in Wolfsburg. The cooperation with Volkswagenwerk GmbH led to an array of improvements to details on the VW Beetle, of which one million were built by 1955 alone. In 1949, Volkswagenwerk GmbH was placed under trusteeship of the Federal Republic of Germany and the state of Lower Saxony.

Ferry Porsche (right) in conversation with General Director of the Volkswagenwerk plant Heinrich Nordhoff.

1949

At the beginning of the year, a central office was set up in Salzburg to manage imports, sales and customer service for the Volkswagen in Austria. On 14 May, Porsche-Salzburg Ges.m.b.H. was founded by Anton Piëch and his wife Louise Piëch.

The first Type 356 cabriolet, equipped with a body from the Swiss bodywork specialist Beutler, was finished in February. Together with the first Type 356 coupe, Porsche presented the sports cars at the Geneva Motor Show on 17 March.

The Porsche booth at the 1949 Geneva Motor Show.

Dutch car dealer Ben Pon became the world's first official importer of Porsche sports cars.

The denazification of Ferdinand Porsche in Germany took place in summer 1949. Despite membership of the National Socialist party, the Central Judgement Chamber for North Württemberg classified him as "not compromised" as he had not held any political or organisational party office.

Due to cramped conditions in Gmünd, Porsche Konstruktionen GmbH was striving for Porsche to return to the car manufacturing hub of Stuttgart. However, as the plant at Spitalwaldstrasse 2 was still occupied by the US military, an office and small workshop were set up at Villa Porsche on Feuerbacher Weg in Stuttgart.

On 26 November, the Stuttgart-based Porsche Konstruktionen GmbH was established alongside Dr. Ing. h.c. F. Porsche KG. It was managed by Ferry Porsche

Albert Prinzing played a key role in the return of Porsche KG to Stuttgart.

and his schoolfriend Albert Prinzing. A short while later, Porsche rented a 600-square-metre hall from Stuttgart-Zuffenhausen-based body shop Reutter & Co. GmbH and began preparing to produce the Type 356. In return, the Reutter body shop received an order to make 500 steel body shells.

Once Italian industrialist Piero Dusio opted not to produce the Type 323 Porsche tractor, Porsche GmbH began working on the construction and testing of the Type 313 tractor, complete with air-cooled two-cylinder diesel engine. The finished design was acquired by the Göppingen based agricultural machinery manufacturer Allgaier in November. Between 1949 and 1957, over 25,000 units of this car were produced.

Draft of the tractor, March 1946.

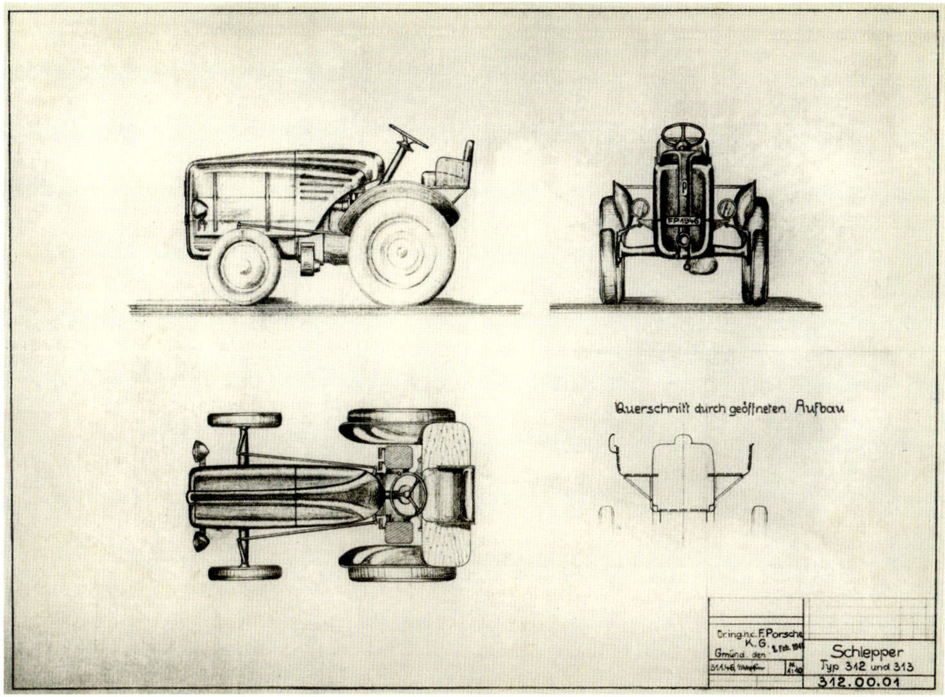

Stirling Moss at the Le Mans start of the 1958 Buenos Aires 1,000-kilometre race.

1950 — PORSCHE WORLDWIDE — 1959

"If I can't sell at least five cars per week, I'm not interested."

Maximilian E. Hoffman

1950

The move to the Württemberg state capital saw the end of the Austrian Porsche Konstruktionen GmbH. The company began trading under the name Dr. Ing. h.c. F. Porsche KG once again. The Stuttgart-based Porsche Konstruktionen GmbH was liquidated two years later.

From March onwards, the first Porsche sports cars were produced in Stuttgart and sold for DM 10,200 for the coupe edition. Walter Glöckler of Frankfurt became the first German Porsche dealer.

In spring, Albert Prinzing presented VW wholesalers and importers at the Wolfsburg Volkswagen plant with a Type 365 coupe and cabriolet produced in Swabia. He returned to Zuffenhausen with 37 orders and ensured production start-up with advance payments worth 200,000 Deutschmarks. By the end of the year, 369 Porsche sports cars were built in collaboration with the Stuttgart body-maker Reutter. Before that, between 1948 and 1950, at the Gmünd plant – in addition to "No. 1" – 52 cars of the Porsche Type 356/2 with light alloy bodies were built.

Completion of the first Porsche 356 built in Stuttgart on 6 April 1950.

The Gmünd engineering office made a full return to Stuttgart in July. As the former Porsche site was still occupied by American forces, the design engineers initially worked at temporary premises. This was where they began designing the 1.5-litre engine (Type 502), which would prove crucial to the future direction of Porsche, in late July.

Ferry Porsche (bottom right) with his father Ferdinand Porsche in the Porsche engineering office in Stuttgart-Zuffenhausen.

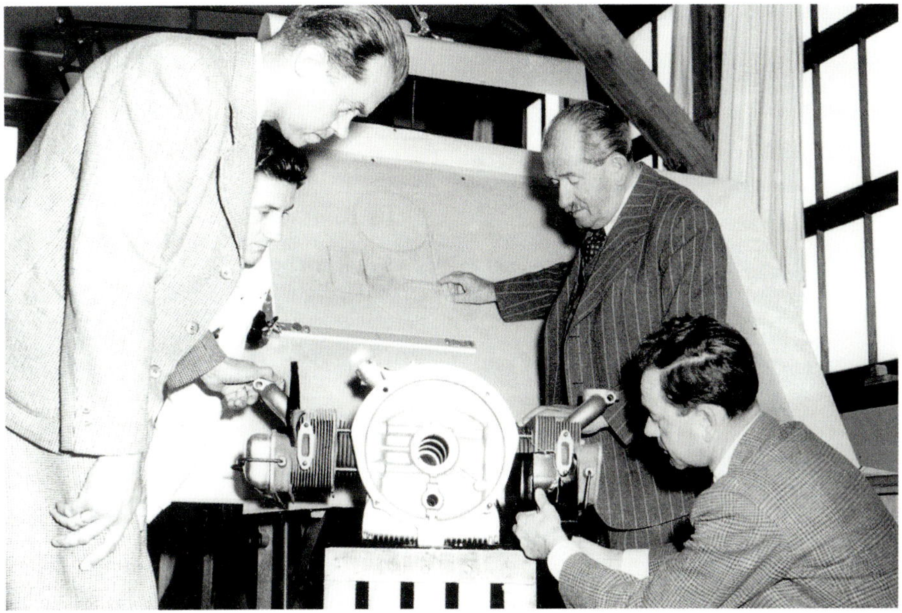

In an encounter arranged by Swiss journalist Max Troesch, a decisive meeting took place between Professor Ferdinand Porsche and American car importer Maximilian E. Hoffman in October at the Paris Motor Show. As a representative of various European car brands, Hoffman had his own network of dealerships on the US East Coast and went on to present the Porsche 356 in New York that same year.

Ferry Porsche (left) and the US importer Maximilian E. Hoffman in New York.

1951

Ferdinand Porsche in 1950.

Company founder and automotive design engineer Professor Ferdinand Porsche died on 30 January, aged 75. His final resting place is in the burial chapel on the family's estate in Zell am See.

At the Geneva Motor Show on 14 March, Porsche KG reached an official agreement with AMAG on the export of Porsche sports cars to Switzerland, supplemented by a general agency agreement on 5 April.

Motorsports had become increasingly important to the fledgling car manufacturer Porsche. Following Otto Mathé's success in the International Alpine Rally the previous year, Porsche attracted international attention with a class win at the 24 Hours of Le Mans in June. French Porsche importer and general agent Auguste Veuillet and Edmond Mouche won the 1.1-litre rating in a Porsche 356 SL. The class win achieved by the team of Paul von Guilleaume and Heinrich Graf von der Mühle in the Liège-Rome-Liège Rally in August also achieved notoriety.

The 72-hour record journey of a light-metal Porsche 356 coupe drew huge attention at the Paris Motor Show in autumn. On the Montlhéry racing circuit south of Paris, a racing driver team set 11 international class records and a new world record with an average speed of 152.34 km/h.

Auguste Veuillet and Edmond Mouche picked up the first class win at Le Mans for Porsche.

The 356 SL during its 72-hour record drive in Montlhéry, France.

Porsche created its first four-seater, based on the 356. However, the Type 530 never made it beyond the prototype stage.

1951 proved a hugely successful year for Porsche KG, as the company produced a total of 1,364 cars in the 356 series. At the end of the year, the Porsche workforce had grown to 214, and sales were more than 11 million Deutschmarks. The lightweight America Roadster (Type 540) was set to join the coupe and cabriolet in the company's range the following year. Produced in a small series, this model was aimed at American customers with motorsports ambitions.

The Type 530, the first attempt at a four-seater.

Porsche 356 America Roadster on the home straight at the airfield race in Chanute, USA, in 1953.

1952

The beginning of the year saw the appearance of the first edition of the Porsche customer magazine Christophorus, the "magazine for the friends of Porsche". The editor-in-chief was renowned journalist and racing driver Richard von Frankenberg. Art direction was handled by Stuttgart-based graphic designer Erich Strenger.

Published by Porsche since 1952, Christophorus was the first customer magazine of its kind.

When Ferry Porsche took a trip to the US in 1951, importer Max Hoffman asked for a quality seal with a reference to "Stuttgart or similar" as an identifying trademark that would indicate the cars' European origin to American customers. After the trip, design engineer Franz Xaver Reimspieß drew the first sketch of the Porsche emblem in early 1952. Its aim was to express identity, quality and dynamism in one single image. Once registered with the German Patent Office, the Porsche emblem appeared on the steering wheel cap for the first time, again in 1952. From November 1954, it adorned the bonnet handle of the Porsche 356 Speedster. Today, it is an indispensable identifying feature of cars from Zuffenhausen.

Original drawing of the Porsche crest by Franz Xaver Reimspieß of the year 1952.

The Porsche Club Hohensyburg, later renamed the Porsche Club Westfalen, was founded in May as the first-ever official association of individual Porsche drivers.

In June, Porsche KG received a development order from American car manufacturer Studebaker. During the two-year cooperation, Porsche developed a modern four-door saloon in the shape of the Type 542. However, due to the American customer's lack of funds, it did not make it to series production.

The Porsche Type 542 (Studebaker) outside Solitude Palace.

In July, production began on a fully synchronised Porsche gearbox with ring synchronisation (Type 519) following its presentation at the Geneva Motor Show. By 1968, over a million synchronised gearboxes would go on to be manufactured under licence worldwide.

The Porsche gearbox with ring synchronisation was intended to set the trend for automotive engineering.

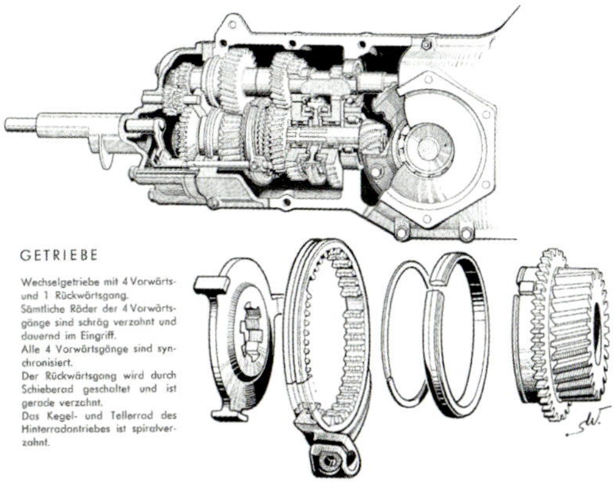

The Porsche 356 proved its reliability in numerous racing events and won the German Sports Car Championship. At the International Eifel Race on Nürburgring, a private Glöckler-Porsche won right away, while Auguste Veuillet and Edmond Mouche set a new class record at the 24 Hours of Le Mans. The Liège-Rome-Liège endurance rally was also dominated by Porsche – with five Type 356 cars placing among the top ten.

In 1952, the Porsche 356 proved its reliability and racing prowess at prestigious racing events.

In November, automotive manufacture began in the new Werk 2 main hall, located next to the Reutter company. At the end of the year, company management and the design and sales departments moved in. The building was designed by Rolf Gutbrod, a freelance architect based in Stuttgart since 1946 who would later achieve global recognition.

The vehicle assembly hall at Werk 2 shortly after its completion in 1952.

1953

At the XX. Mille Miglia in April 1953, a total of 18 Porsche sports cars from six different countries took part. All reached the destination in Brescia after 1,600 kilometres, racking up wins in the 1.3 litre and 1.5-litre capacity classes.

After the Mille Miglia Porsche used new track records for advertisement.

For the first time ever, two Porsche 550 sports cars lined up in Le Mans with driver teams Helm Glöckler and Hans Herrmann, and Richard von Frankenberg and Paul Frère respectively. With an average speed of 138.8 km/h, the racing car with its closed structure not only secured the two top spots in the 1.5-litre capacity class – it also beat the previous class record by 16 km/h.

For aerodynamic reasons, the two 550 Spyder at the 24 Hours of Le Mans started with an attached roof.

Early August saw the first outings of the four-cam-shaft engine Type 547, designed by Porsche engineer Ernst Fuhrmann, on the Nürburgring racetrack in training for the sports car classification race. The engine was presented to the public for the first time

The Porsche 550 Spyder with the Type 547 Carrera engine designed as a racing engine.

at the Paris Motor Show in October. In addition to the Porsche 550 racing car, the Carrera engine would also go on to be used in subsequent Carrera versions of the 356.

Porsche KG scored numerous overall and class wins in the 1953 racing season.

Richard Trenkel and Hans Herrmann, crowned an outstanding year for Porsche in racing by taking German Sports Car Championship titles in the 1.1-litre and 1.5-litre classes respectively.

Porsche design engineers handled a range of different orders, in addition to continuing to develop its sports cars. In autumn, for instance, a VW prototype with a self-supporting structure and a shortened wheelbase, the Type 534, was presented to Heinrich Nordhoff, the General Director of Volkswagen.

December saw the beginning of the development of the Type 597 "Jagdwagen", designed as an amphibious all-wheel-drive vehicle as part of a Bundeswehr (German armed forces) tender. However, for reasons pertaining to employment policy, the public order was awarded to the Auto Union.

Driving tests with Type 597.

1954

Since 1954, the Porsche emblem has graced the front of every Porsche car. The trademark symbol was designed back in 1952 by Ferry Porsche and Franz Xaver Reimspieß. At the suggestion of foreign dealers and customers, they combined the Porsche logo created in 1951 with the Stuttgart "Rössle" horse and the Württemberg state coat of arms.

Porsche enthusiasts took part in racing events all over the world in 1954, making Porsche synonymous with sporting excellence, and scored 206 rally and 214 race successes. Carl-Gunnar topped the overall placings in a Type 356 at the Swedish Rally to the Midnight Sun. At the 12 Hours of Reims, the team of Helmut Polensky and Richard von Frankenberg and of Auguste Veuillet and Jacky Olivier took the top two class spots in their Porsche 550 Spyder cars. In addition to racking up four wins at Nürburgring, the Liège-Rome-Liège rally once again became a festival of Porsche, with the overall winner and six of the top 11 cars coming from the Porsche stable. Other highlights included Hans Hermann's German Sports Car Championship title, a win in the sports car category at the Tour de France by Claude Storez and Herbert Linge and class wins at the Mille Miglia (Herrmann/Linge) and the Carrera Panamericana (Herrmann/Linge). This success against strong opponents earned the four-camshaft engine the nickname Carrera, which means "race" in Spanish. This sobriquet was soon introduced into the designated name for 356 cars with the four-camshaft engine. The Porsche 911 with its six-cylinder boxer engine also bears the name with pride today.

Private drivers and factory teams alike triumphed in Porsche cars.

Although the Bundeswehr's test of the Porsche Type 597 was a success, a government committee decided against the all-wheel-drive off-road vehicle from Zuffenhausen at the end of the year. Since the Porsche factory was at full capacity with the production of sports cars, a civilian version of the Type 597 was ultimately decided against, and so a pre-series of 71 units of the Porsche "Jagdwagen" was all that remained.

The Porsche Type 597 "Jagdwagen" offered outstanding off-road capability.

With almost 500 employees, Porsche KG produced a total of 1,853 cars. From September, these included the first 15 sports cars known as the Type 356 Speedster, a lightweight version especially designed for the US market – an American sales success at a basic price of around 3,000 US dollars.

The 356 Speedster boasted a powerful four-cylinder boxer engine with 40 kW (55 PS).

1955

Porsche once again displayed its prowess at numerous endurance races. At the Mille Miglia, all three Porsche 550 Spyder cars reached the finish line, securing first, third and fourth places in class. At Rally Sweden leading across the Arctic Circle, Type 356 sports cars secured the top four spots in the overall ranking. On the circuit, Porsche shone with results such as another Le Mans class win and a German Championship win in the 1.5-litre sports car class by Richard von Frankenberg.

The Seidel/Glöckler team achieved a class win at the Mille Miglia in 1955 in the Porsche 550 Spyder.

Porsche design engineers were commissioned by Volkswagen to design a small rear-drive car known as Type 672. A boxer six-cylinder underfloor engine was designed as a drive for the prototype.

Technical drawing of the small car project prototype Type 672, 7 July 1955.

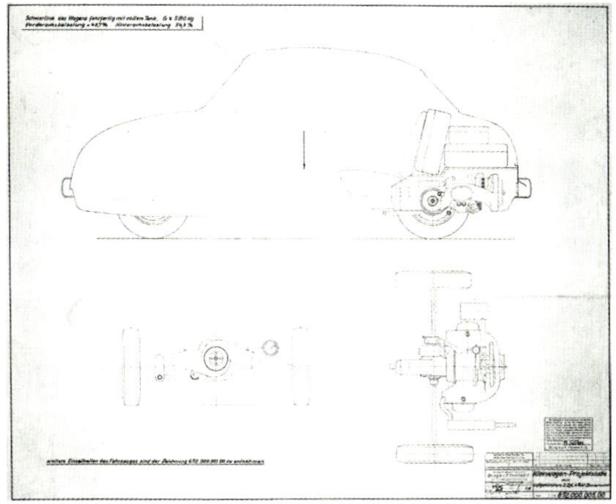

The Porsche 356 A equipped with the new 1.6-litre engine series went on display at the International Motor Show in Frankfurt am Main in September. Outwardly, the technically improved A series can be

The Porsche 356 A Coupé as presented at the 1955 International Motor Show in Frankfurt am Main.

recognised by its curved panoramic windscreen. The top model available was the Porsche 356 A 1500 GS Carrera with a 74 kW (100 PS) racing engine.

Exports played an increasingly important role for Porsche KG, and 1955 saw almost half of annual production sold overseas, with the export of 1,514 sports cars to North America. Dissatisfaction with the customer service being provided by the Hoffman Motor Car Company saw Porsche employee Otto Erich Filius head out on secondment. He set up the first Porsche office in New York and built an independent sales network with the name Porsche of America Corporation (POAC).

Porsche cars were shipped overseas from an early stage.

Aerial view of the Porsche KG factory site and body manufacturer Reutter.

On 1 December, the US Army released the main Porsche plant on Spitalwaldstrasse. From March 1956, company management, the design, test and development divisions, sales and repairs would all begin work there.

1956

Porsche-Diesel-Motorenbau GmbH Friedrichshafen a.B., part of the Mannesmann Group, was founded in Friedrichshafen on Lake Constance on 1 January. In a deal brokered by Albert Prinzing, the Allgaier share capital from Mannesmann AG was also absorbed and transferred to the newly established company. In autumn of that year, production of the tractors and stationary engines now branded Porsche-Diesel continued in Manzell.

In time for the company's 25th anniversary, the 10,000th Porsche sports car – a Porsche 356 A Coupé – was produced on 16 March. By this point, Porsche had chalked up 400 racing victories.

Wolfgang Porsche alongside the 10,000th Porsche.

The 1000-kilometre race, at Nürburgring held on 27 May became another Porsche triumph as the driver teams Maglioli/von Trips and Herrmann/von Frankenberg achieved a double victory in

Major triumphs at the 1,000-kilometre race at Nürburgring.

the 1.5-litre racing class with the 550 A Spyder. A Porsche 356 A Carrera won the two-litre GT class, while a Porsche 550 Spyder triumphed among the series sports cars up to 1,500 cc. On 29 July, the 24 Hours of Le Mans ended in a class win for Porsche for the sixth time in a row.

The Targa Florio on 10 June, Umberto Maglioli reached the finish line in a Porsche 550 A Spyder as the overall winner. This saw Porsche secure worldwide recognition once again, as for the first time a driver in the small car class with a capacity of less than two litres succeeded in beating the much stronger competition from Ferrari and Maserati.

In 1956, Umberto Maglioli picked up the first overall win of Porsche in a major international race in a 550 A Spyder.

Over 75 per cent of 356 A sales were exports.

In the 1956 financial year, Porsche achieved a new sales record with the 356 A coupes, cabriolets and Speedster. Of the sports cars available with five different engines, 4,269 units left the Zuffenhausen factory premises – more than 75 per cent of them exports.

At the end of the year, Porsche introduced an occupational retirement scheme in the form of a monthly pension from the age of 65. A Porsche foundation was established to offer financial subsidies for employees who ended up in financial difficulty through no fault of their own.

1957

The motorsport season began with huge success for the Zuffenhausen racing division. The American private drivers Bunker/Wallace and Kunstle/Miles succeeded in picking up a double victory in the class for sports cars up to 1,500 cc in a Porsche 550 A Spyder in the 12 Hours of Sebring, part of the Sports Car World Championship.

Porsche made a successful start to the 1957 racing season with an overall win in the 12 Hours of Sebring.

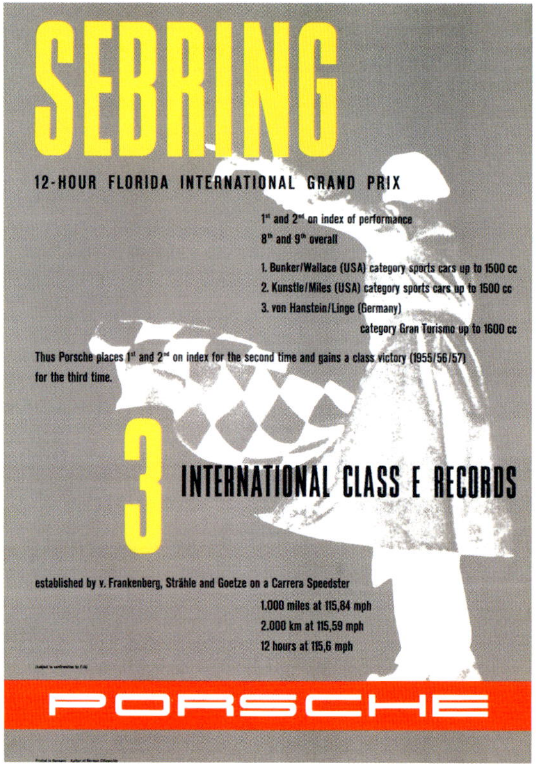

On the high-speed track at Monza, Rolf Goetze, Paul Ernst Strähle and Richard von Frankenberg broke three international driving records over 1,000 miles, 2,000 kilometres and 12 hours in a Porsche 356 A Carrera Speedster.

Richard von Frankenberg in the Porsche 356 A Carrera Speedster in Monza, 1957.

From April, Porsche-Diesel-Motorenbau GmbH continued the series production of agricultural machinery and engines in four newly built production facilities. The new Porsche-Diesel sales portfolio was presented in October and included the basic types Junior, Standard, Super and Master, covering a performance range from 14 PS to 50 PS.

A Porsche 550 A Spyder works car took part in the last Mille Miglia in May and picked up a class win and fifth place overall with Umberto Maglioli at the wheel. In the GT class up to 1,600 cc, Paul Ernst Strähle and Herbert Linge took the win in Strähle's Porsche 356 A Carrera GT.

Advertisement for Porsche-Diesel-Motorenbau GmbH in Friedrichshafen on Lake Constance.

Umberto Maglioli at the Mille Miglia finish line on 12 May 1957.

The Porsche 550 A Spyder at the 1,000-kilometre race at Nürburgring.

After a quadruple victory in the 1,000-kilometre race at Nürburgring and the class win by private drivers Hugus/de Beaufort at the 24 Hours of Le Mans, Porsche started in Formula 2 for the first time on 4 August at the Nürburgring Grand Prix. Edgar Barth won the race with the Porsche 550 A Spyder.

From September, the variety of models in the Porsche sales portfolio would expand to include the Porsche 356 A Cabriolet, complete with hardtop as a removable roof attachment.

356 A Cabriolet with hardtop.

1958

By now, 142 German Porsche dealerships and another 128 sales partners elsewhere in Europe were selling cars from Zuffenhausen.

A look at Porsche Werk 2 in 1958.

August saw the Porsche 356 A Speedster replaced by the 356 A Convertible D, assembled by Heilbronn-based bodywork specialists Drauz. On the outside, the successor to the Speedster was recognisable thanks to its taller, reinforced windscreen and glass side windows.

The 356 A Convertible D was made by Heilbronn-based bodywork specialists Drauz.

At the 1,000-kilometre race in Buenos Aires on 26 January, racing drivers Stirling Moss and Jean Behra took third place behind two three-litre Ferraris in a Porsche 550 A Spyder and won the two-litre sports car class decisively.

The air-cooled Porsche boxer engine Type 729 was also used as a drive for racing boats.

The Porsche Type 729 boat engine was installed in a sports boat for testing and demonstration purposes in spring. The air-cooled 52 PS drive was used by various boat manufacturers in subsequent years but was never able to establish itself on the boat engine market.

In Formula 2, Frenchman Jean Behra beat the world elite in this racing category in Reims in the Porsche 718 RSK, which was actually designed for the sports car class. Further success finally prompted company boss Ferry Porsche to agree to the development of a new Formula 2 Monoposto.

Jean Behra in a 718 RSK "Mittellenker" (centre steer) at the Reims Grand Prix in 1958.

In addition to such titles as German Champion in the sports car class with Jean Behra and European-Champion in the mountains ranking with Wolfgang Graf Berghe von Trips, Porsche secured the runner-up spot in the World Championships in the sports car manufacturer classification in 1958.

Porsche secured the European Hill Climb Championship for the first time in 1958.

1959

With the 356 B series – known internally as the "T 5" – Porsche presented a completely overhauled version of the successful 356 model, which has since been manufactured almost 29,000 times, at the Frankfurt International Motor Show in September. A Roadster version replaced the Convertible D and offered a lower-priced addition to the sales portfolio. A total of 6,450 units were sold in 1959.

Porsche presented the 356 B as the successor to the 356 A at the 1959 International Motor Show (IAA) in Frankfurt am Main.

During the German Industrial Fair in Hanover, Porsche presented a complete range of aircraft engines. The performance spectrum ranged from Type 678/1 with 65 PS to Type 678/4 with 75 PS starting capacity.

The Porsche aircraft engine Type 678 in the RW-3 "Multoplane" aircraft.

1959 was another year marked by sporting success: at the 12 Hours of Sebring, the Porsche 718 RSK took third, fourth and fifth place overall and won the 1.5- and two-litre capacity classes. A subsequent victory in Germany's Grand Mountain Prize and second place overall with a class win in the English Tourist Trophy were just as decisive. Thanks to Paul Ernst Strähle, the Liège-Yugoslavia-Liège Rally became another victory for Porsche. The track record was further enhanced by the German Sports Car Championship title picked up by Heini Walter and the European Hill Climb Championship secured by Targa Florio winner Edgar Barth.

Joakim Bonnier starting the 12 Hours of Sebring in a 718 RSK.

In August, the team led by Ferry Porsche's son Ferdinand Alexander "F. A." Porsche began work on a four-seater Type 695 sports car. On the extended basis of a Porsche 356, a prototype was created as a "model car" for the "T 7" programme (Type 754). The distinctive front section could also be found in the later Porsche 911.

Towards the end of the year, Maximilian E. Hoffman handed over representation – with the exception of the American East Coast – to the Porsche of America Corporation (POAC), which from then on was solely responsible for sales, service and marketing in the United States on behalf of Porsche KG.

Ferry (right) and Ferdinand Alexander Porsche in the Porsche design studio.

F. A. Porsche and his team with the 1963 Porsche 901. From left to right: Heinrich Klie, Heinz Unger, Ernst Bolt, Horst Handke, Hans Ploch and Hans Springmann.

1960 — ICON PORSCHE 911 — 1969

"Good design should be honest."

Ferdinand Alexander Porsche

1930 1940 1950 1960s 1980 1990 2000 2010 2020

1960

Different versions of the Porsche Type 728.

In spring, the Type 728 (EA 53), designed as a commissioned work, was presented to the Volkswagen board. Conceived as a potential successor to the VW "Beetle" in notchback and hatchback designs, the project didn't, however, make it past the testing stage.

Winners' poster after the hat-trick at the European Hill Climb Championship.

With the 718 RS 60 Spyder introduced for the 1960 racing season, Porsche began a new chapter of success in endurance racing. In addition to overall victories in the Targa Florio (Bonnier/Herrmann) and the 12 Hours of Sebring (Gendebien/Herrmann), the Swiss Heini Walter won the European Hill Climb Championship for Porsche for the third time in a row.

In Formula 2, Porsche achieved an impressive triple victory in the 150-mile race in Aintree, Great Britain, with the 718/2. At the end of the season, Porsche won the manufacturers' world championship, encouraging Managing Director Ferry Porsche to finally agree to enter Formula 1.

Edgar Barth (front) and John Surtees in two Porsche 718/2 at Werk 1 in Zuffenhausen.

In September, the Porsche Museum opened in Hall 3 of the Porsche Werk 2 in Zuffenhausen. Alongside the Sascha racing car designed by Professor Ferdinand Porsche, exhibits included the Porsche 356 "No. 1" Roadster and the Type 360 Cisitalia racing car.

The Austro-Daimler ADS R "Sascha" on the Porsche factory site.

At the instigation of Herbert Linge, the new Porsche development centre was built in Weissach.

On 2 December, Porsche KG acquired a plot of land, initially 38 hectares, in the districts of Weissach and Flacht, 25 kilometres from Zuffenhausen, as an area for a future test site and to expand the factory facilities.

In 1960, Dr. Ing. h.c. F. Porsche KG saw sales exceed 100 million Deutschmarks for the first time ever, producing 7,499 cars. The total number of Porsche sports cars manufactured to date was 39,071 cars. A customer service organisation of 653 authorised Porsche workshops worldwide was available.

From as early as 1960, Porsche KG employees received sick pay and a full monthly salary as a Christmas bonus – even without statutory regulations.

1961

Porsche entered Formula 1, limited to a 1.5-litre capacity, with the four-cylinder Type 718/2 Monoposto. At the Grand Prix of Reims, Monza and Watkins Glen, works driver Dan Gurney was second on the podium. Overall, the first year of Porsche involvement in Formula 1 saw Dan Gurney placed third in the world championship. The development of a new eight-cylinder engine was expedited for 1962 accordingly.

Dan Gurney in a Porsche 718 Formula 2, training for Formula 1 racing in Monaco, 1961.

The sportiest version of the 356 model series was the Porsche 356 B 1600 GS Carrera GTL Abarth, built the previous year in a limited edition of 20 cars. Racing drivers such as Hans Herrmann, Herbert Linge, Edgar Barth and Paul Ernst Strähle won the world championship title for GT cars up to 2,000 cc in 1961 at the Abarth GTL for Porsche. Other international titles in 1961 included Heini Walter's fourth win in the European Hill Climb Championship and Hans-Joachim Walter's European Rally Championship victory.

Bob Holbert and Don Wester picked up a class win with the 356 B 1600 GS Carrera GTL Abarth at the 12 Hours of Sebring.

In September, the Porsche 356 B underwent a major facelift under the internal designation T6. Many details of the body had been optimised, now making refuelling possible via a tank flap in the front right wing. An electric steel sunroof was made available for the first time as a special feature.

The Porsche 356 B Cabriolet (model year 1962) on the Zuffenhausen site.

On 16 October, Ferry Porsche broke ground on the new building site in Weissach. Among other things, plans for the future test site included a "skid pad" for the suspension tests carried out at Malmsheim Airport since 1954.

Test drives on the new skid pad at the Porsche development centre in Weissach.

The Porsche Type 754 was a forerunner of the later Type 901, or 911.

In the course of winter test drives with the four-seater Porsche 695 prototype, testing of a new 745 six-cylinder boxer engine with an axial fan began in December.

1962

In spring, Porsche offered its customers a new top sports car model: with a sales price of DM 24,850, the Porsche 356 B 2000 GS Carrera 2 was aimed at a clientele that was both well heeled and performance-hungry. Originally developed by Ernst Fuhrmann, the two-litre Carrera engine now offered 96 kW (130 PS), and up to 114 kW (155 PS) in the GT version with sports exhaust. Racing success was very much in play and the Carrera 2 racked up achievements such as a class win at Rallye Monte-Carlo.

Porsche 356 B 2000 GS Carrera 2 de Luxe Coupé.

The 50,000th Porsche rolled off the assembly line on 3 April. At this time, the daily production volume was around 50 Porsche sports cars.

Celebrating the 50,000th Porsche at Werk 2 in Stuttgart-Zuffenhausen.

After careful consideration, Ferry Porsche decided against the further development of the four-seater prototype 754 in favour of a hatchback coupe with a 2+2 seating arrangement with the programme designation "T 8". The future Porsche 356 successor was being developed under the project name Type 901, incorporating numerous detailed solutions of the 695. A new Type 821 six-cylinder engine with dry sump lubrication was based on the Type 745 drive intended for the "T 7". The team led by Ferry Porsche's son F. A. Porsche took charge of the style of the 901, creating a modern classic with timeless lines.

The newly designed six-cylinder Type 821 engine powered the Porsche Type 901.

The Type 804 saw Porsche introduce a specially designed car for Formula 1 for the very first time. It helped Dan Gurney secure the French Grand Prix in Rouen on 8 July and bring home the first Formula 1 victory to Zuffenhausen. Just one week later, Gurney triumphed ahead of Joakim Bonnier on the Solitude racetrack near Stuttgart – so the air-cooled Porsche 804 enjoyed a double triumph before a home crowd.

The first stage of the new test site in Weissach was completed and entered operation towards the end of the year. The new test track was used for chassis tuning; the new skid pad had an outer lane ring with a diameter of 190 metres and two inner tracks with a radius of 30 and 20 metres respectively.

Dan Gurney won the French Grand Prix in a Porsche 804.

The test tracks for chassis tuning at the Porsche test site in Weissach were expanded.

1963

Porsche-Diesel-Motorenbau GmbH ceased its operations on 1 May. The shares were transferred to the newly founded Porsche-Diesel-Renault-Vertriebs GmbH, which also took over the sales network in June. The production of Porsche diesel tractors came to an end on 15 July. A total of around 120,000 tractors had been produced and sold under the Porsche name.

The Porsche-Diesel Master tractor.

On 12 September, a prototype of the new Porsche 901 was presented for the first time at the International Motor Show (IAA) in Frankfurt. In addition to a six-cylinder engine with overhead camshafts, the successor to the by now already legendary Porsche 356 had a chassis with a semi-trailing arm rear axle and a self-supporting body.

Introducing the new Porsche Type 901 at the Frankfurt IAA Motor Show in September 1963.

The IAA booth also showcased the new Porsche 356 C model series, which, in addition to many technical improvements, was equipped with a four-disc brake system and built in series parallel to the later 911 until September 1965.

The 356 C is the last version of the 356 model series, in production since 1948.

Starting at the Targa Florio in a Porsche Type 718 GTR Coupé, the driver team of Joakim Bonnier and Carlo Abate not only brought back the fourth Porsche overall victory to Zuffenhausen after 1956, 1959 and 1960, but also set a new course record with an average speed of 104 km/h.

Joakim Bonnier and Carlo Abate won the 1963 Targa Florio with a new speed record.

With the eight-cylinder Porsche Type 718 W-RS Spyder, Edgar Barth won six out of seven championship races and became the outstanding European Hill Climb Champion.

During the transition to the new Type 901, the Stuttgart bodywork factory Reutter in Zuffenhausen was sold to Porsche. As of 1 December, around 950 employees changed employers, taking into account their years of service. The new owner continued the chassis works under the existing name. On 1 March 1964, the name was changed to Karosserieweerk Porsche GmbH, before being merged into Porsche KG in 1967.

Successful on the Hills.

Bodyshop factory Reutter in Zuffenhausen.

1964

As the first Porsche, with a fibreglass-reinforced plastic body, the Porsche 904 Carrera GTS Coupé turned heads with its design by F. A. Porsche and its driving performance alike. It wasn't long before the racing car scored its first wins. The mid-engined sports car saw success in races like the Targa Florio (Pucci/Davis) and secured second place at the 1965 Rallye Monte-Carlo (Böhringer/Wütherich).

The illuminated glass-fibre-reinforced body of the Porsche 904 Carrera GTS.

After a test series of 13 prototypes of the 901 was successfully put through its paces, the first Porsche Type 901 in series rolled off the production line on 14 September. In October, the model was presented, ready for series production, at the Paris Motor Show. As French manufacturer Peugeot already had the name rights to three-digit sequences of numbers with a zero in the middle, the Stuttgart-based company had to rename the model series: the 911 was

Porsche employee Thora Hornung with a Porsche 901 Coupé at the Porsche Werk 2 site in Zuffenhausen.

born. By this time, a few cars with the 901 designation had already been produced. By the end of the year, around five Porsche 911 cars were leaving the factory each day.

In autumn, the Porsche of America Corporation (POAC) took over the import and sales of Porsche cars on the US East Coast, which had until then remained the responsibility of Maximilian E. Hoffman.

At the end of the year, Dr. Ing. h.c. Porsche KG and Porsche Karosseriewerk GmbH together employed a total of 2,415 people, producing 10,368 cars and achieving sales of 156 million Deutschmarks.

Aerial view of Stuttgart Zuffenhausen with Schwieberdinger Straße in the centre.

1965

The Porsche 911 proved its prowess as a racing car at Rallye Monte-Carlo – the toughest in history thanks to difficult weather conditions. After 4,600 kilometres of snow, only 22 of the 237 cars that started the race made it to the finish line. Herbert Linge and Peter Falk ended the race in a respectable fifth place in the overall rankings. Eugen Böhringer and Rolf Wütherich made second place in a Porsche 904 Carrera GTS.

The Porsche works team before leaving for the 1965 Rallye Monte-Carlo at Porsche Werk 1.

After the sales department, called for an affordable entry-level model below the Porsche 911 Coupé costing 21,900 Deutschmarks, the four-cylinder Porsche 912 was introduced in April. Equipped with the 356 C 1600 SC engine throttled to 66 kW (90 PS), the 912 Coupé, already available at a starting price of 16,250 Deutschmarks, quickly became a hit.

Manufacturing the Porsche 911 2.0 Coupé and 912 Coupé at Porsche Werk 2 in Stuttgart-Zuffenhausen, 1965.

The second body shape for the 911 2.0 is the Targa with fixed roll bar.

The world's first safety cabriolet, the Porsche 911 2.0 Targa, was presented to the public at the Frankfurt Motor Show in September. With its fixed roll bar, the removable folding roof and lowerable rear window, the new version took into account American customers' higher safety requirements. Production of the Porsche 911 Targa began in December 1966.

At the 1965 Targa Florio, Colin Davis and Gerhard Mitter took second place overall in a 904/8 "Kangaroo".

At 12 Hours of Sebring, four of the top ten cars that placed overall were a Porsche 904 Carrera GTS. Another surprise came when Gerhard Mitter and Colin Davis placed second overall in the Targa Florio in the Porsche 904/8 "Kangaroo".

On 30 September, it was Porsche Press Director and Racing Manager Huschke von Hanstein who brought home two world records for the quarter-mile and 500-metre distances to Zuffenhausen in a 488 kilogrammes Porsche 906 Spyder "Ollon-Villars" Bergspyder at Hockenheimring.

Porsche achieved two new world records on Hockenheimring.

In 1965, the Technical University of Vienna awarded Ferry Porsche an honorary doctorate. From then on, he was able to use the title "Dr. tech. E.h.".

Ferry Porsche received an honorary doctorate in Vienna for his services to automotive engineering on 9 November 1965.

1966

After the 356 model series was officially discontinued the previous year, a final special series of ten Porsche 356 C Cabriolet for the Dutch Rijkspolitie police service rolled off the production line on 26 May. From 1950 to 1966, Stuttgart-Zuffenhausen produced a total of 77,766 of the classic cars bearing the figures "356".

With the new Porsche 911, Porsche racked up win after win. At Rallye Monte-Carlo, the four Porsche racing cars that started the race took the top four places in the GT category.

The Porsche 911 2.0 Coupé showed its sporting prowess at the 1966 Rallye Monte-Carlo.

The Porsche 906 Carrera 6 – a pure racing sports car with a tubular space frame and aerodynamic plastic body – made its motorsporting debut at the opening race of the manufacturers' world championship, the 24 Hours of Daytona. At the end

Hans Herrmann and Herbert Linge picked up a class win in a 906 Carrera 6 in Daytona in 1966.

The sports car class with up to two-litre capacity was dominated in 1966 by the Porsche 906 Carrera 6.

of the tournament, three of the Porsche GT prototypes were among the top eight placements in the overall ranking.

At the 12 Hours of Sebring, the 906 Carrera 6 went toe to toe with a raft of American seven-litre racing cars and scored three class wins and fourth place in the overall ranking. In the two-litre sports car class, the racing car proved unbeatable, chalking up the Targa Florio won by the driver team Müller/Mairesse as another success. Le Mans saw three 906 Carrera 6 cars with their long tail line up to start the race and go on to occupy the top three places in their class.

In autumn, Porsche presented the 911 S, a performance-enhanced version of the sports car. Outwardly, the 118 kW (160 PS) Porsche is recognisable by its forged Fuchs cloverleaf alloy wheels.

The famous Fuchs rims were installed on the 911 S for the first time in model year 1967.

The 100,000th Porsche, a 912 Targa especially equipped for the autobahn police, was given to the Baden-Württemberg state police at a special ceremony on 21 December.

Ferry Porsche giving the 100,000th Porsche ever made, a 912 Targa, to the Baden-Württemberg state police.

1967

From July, car manufacturer Porsche began offering a semi-automatic four-speed transmission for the 911 and 912 models under the name "Sportomatic". It was now also possible to acquire the Porsche 911 T (Touring). This was a more affordable basic alternative to the classic sports car now known as the 911 L.

In 1967, the "Sportomatic" proved its reliability at the Marathon de la Route at Nürburgring.

At the very start of the season, with the Porsche 906 Carrera 6, Hans Herrmann and Jo Siffert decisively won the two-litre class at 24 Hours of Daytona, achieving fourth place overall. At the 12 Hours of Sebring, the new Porsche 910/6 had reached an outstanding third place in the overall ranking.

With a fleet of six 910 prototypes, the Swabian racing team took part in the Targa Florio in May. The race ended for Porsche with a triumphant triple win, when Rolf Stommelen and Paul Hawkins in their eight-cylinder Porsche 910/8 pipped two 910/6 cars to the finish line. Another Porsche 911 S works car with Bernard Cahier and Jean-Claude Killy claimed the two-litre GT class.

The Porsche team with six Type 910 racing cars at the Targa Florio.

On 28 May, the 1,000-kilometre race at Nürburgring turned out to be one of the greatest racing successes, and Porsche was able to win the long-distance race for the first time. After 44 laps of the race, the Type 910 racing cars ranked in the top four places and the sixth.

A quadruple victory for Porsche at the 1,000-kilometre race at Nürburgring in 1967.

Presented in October, the Porsche 911 R attracted international attention when it set new world records in Monza over 20,000 and 25,000 kilometres as well as 10,000 miles and 72 and 96 hours. A Porsche 911 R with Sportomatic transmission had already proven its reliability during the 84-hour Marathon de la Route at Nürburgring.

The Porsche 911 R breaking records in Monza, 1967.

Porsche became the first European manufacturer to follow American exhaust purification specifications. On behalf of the American Environmental Protection Agency (EPA), Porsche henceforth began testing export vehicles from other manufacturers on its own test bench.

1968

Björn Waldegård and Lars Helmer celebrated a decisive overall win in the 3,400-kilometre Rally Sweden in a Porsche 911 T. Rallye Monte-Carlo ended in first place overall for Vic Elford and David Stone in the Porsche 911 T. Porsche team Pauli Toivonen and Martti Tiukkanen claimed second place.

At the 1968 Rallye Monte-Carlo, Vic Elford and David Stone were overall winners in a 911 T 2.0 Coupé.

The Porsche 907/8 also achieved a stunning racing debut with a triple win at the 24 Hours of Daytona and first place at the 12 Hours of Sebring.

Porsche winners' poster after the success at Daytona.

The Porsche works team won the 1968 Targa Florio with a hat trick. With the victory of Vic Elford and Umberto Maglioli in the 907/8, the Porsche company finally picked up the "Coppa Florio" trophy, which was given place of honour in Ferry Porsche's study.

With Vic Elford and Umberto Maglioli at the wheel of a 907, Porsche won the Targa Florio for the third time running.

In May, Porsche once more repeated a previous year's victory by winning the 1,000-kilometre race at Nürburgring. Jo Siffert and Vic Elford in the 259 kW (350 PS) Porsche 908/01 pipped Hans Herrmann and Rolf Stommelen in the 907/8 to the post.

On 19 May 1968, the Porsche 908 triumphed in the 1,000-kilometre race at Nürburgring.

The Porsche 910/8 Bergspyder weighing just 450 kilogrammes and with 202 kW (275 PS), set new standards in mountain racing. Porsche works driver Gerhard Mitter brought the European Hill Climb Championship back to Zuffenhausen for the third time in a row.

Gerhard Mitter became the European Hill Climb Champion in 1968 while driving a Porsche.

In the sports car segment, the Porsche 911 E, equipped with a mechanical dual-row injection pump, replaced the "L" version in the summer. Now boasting 103 kW (140 PS), the coupe was now available for 21,645 Deutschmark – the Targa version cost around 2,000 more.

1969

In spring, at the Geneva Motor Show, Porsche presented the Type 917 developed under the direction of Ferdinand Piëch for the first time. By the end of April, 25 samples of the car had already been produced as a homologation series. With a 4.5-litre 12- cylinder engine, the 917 was intended to replace the successful 907 and 908 types and cause a sensation in the capacity class of up to five litres.

The 25 units for the homologation of the 917 at the Porsche workshop.

After the Porsche 917 initially failed in its first three outings due to technical defects, the success story of the now legendary Porsche racing car shaped as long-tail (LH) and short-tail (KH), which lasted until 1973, began on 10 August at the 1,000-kilometre race on the Österreichring with a victory for Jo Siffert and Kurt Ahrens.

Jo Siffert and Kurt Ahrens celebrating the first victory of a 917 KH at the Österreichring in 1969.

The most successful sports car of the 1969 season was the Porsche 908/02, designed as an open Spyder due to a rule change. The greatest triumphs of the 600 kilogrammes racing car include successes at

Ferry Porsche (fourth from left) in the circle of racing drivers at the 1969 Targa Florio.

Brands Hatch (triple win), in Monza (triple win), at the Targa Florio (quadruple win) and at Watkins Glen (triple win). At the end of the racing year 1969, Porsche decisively won the manufacturers' world championship.

The Porsche 911 S impressively demonstrated its reliability in rally racing with a double victory at Rallye Monte-Carlo (Waldegård/Helmer and Larrousse/Perramond).

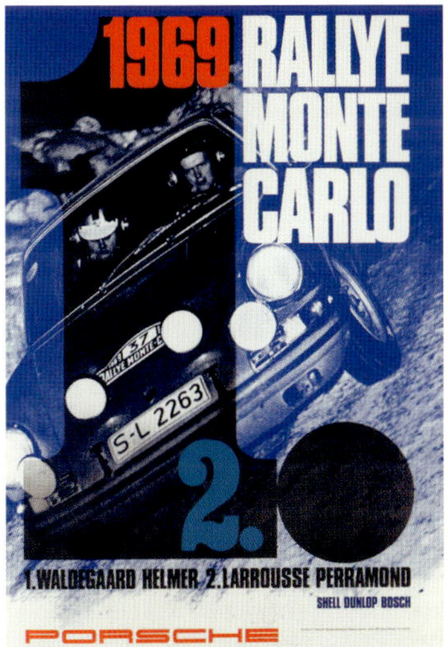

Porsche took the top two spots at the 1969 Rallye Monte-Carlo.

The VW-Porsche-Vertriebsgesellschaft mbH was founded in April 1969 with working capital of five million Deutschmarks. The purpose of the organisation was to sell the new Types 914/4 and 914/6, jointly developed by Porsche and Volkswagen and presented at the Frankfurt IAA Motor Show in autumn 1969.

914/4 (left) and 914/6 at the IAA Motor Show in Frankfurt, 1969.

The Porsche of America Corporation (POAC) was wound up at the end of 1969 and its activities were taken over by the Porsche-Audi division of Volkswagen of America Inc.

Last pit stop before Helmut Marko and Gijs van Lennep win the 1971 24 Hours of Le Mans in a 917 KH.

1970 — OVERALL VICTORIES — 1979

"The biggest risk of my life."

Ferdinand Piëch

1930 1940 1950 1960 1970s 1990 2000 2010 2020

1970

With a basic price of less than DM 12,000, the VW-Porsche 914 became a hit. A total of 13,312 of the four-cylinder version with the 1.7-litre engine were sold in 1970 alone. With a sales price of DM 19,980, the top-of-the-range Porsche 914/6 was equipped with the two-litre six-cylinder engine with 81 kW (110 PS) from the Porsche 911 T.

A mid-engine for the VW-Porsche 914/4.

For the first time in the history of the Rallye Monte-Carlo, Porsche managed to rack up a double win for the third time in a row with the 911 S. Further triumphs followed, securing the Porsche victory in the rally manufacturers' world championship at the end of the year.

Björn Waldegård and Lars Helmer took first place in the overall ranking at the Rallye Monte-Carlo.

Hans Herrmann and Richard Attwood achieved the first overall Porsche win at Le Mans with the 917.

Clinching nine out of the ten possible victories, Porsche once again commandingly won the manufacturers' championship with the Types 917 and 908/03. The series of wins began in Daytona and continued in Brands Hatch, Monza, Spa, on the Nürburgring racetrack, at the Targa Florio, in Le Mans, Watkins Glen and the Österreichring. The performance of Hans Herrmann and Richard Attwood on 14 June 1970 in Le Mans was particularly remarkable. In a 917 KH from Porsche Salzburg, the two drivers picked up an overall win at Le Mans for the Stuttgart car manufacturer for the very first time.

The Targa Florio ended on 3 May 1970 with a double win for Siffert/Redman and Kinnunen/Rodriguez in the Porsche 908/03 Spyder. The ultra-lightweight racing car featured in a total of four races in 1970 and 1971, of which it won three.

Jo Siffert and Brian Redman won the 1970 Targa Florio in the Porsche 908/03.

In September 1970, Ferry Porsche and Louise Piëch, the partners in Dr. Ing. h.c. F. Porsche KG, raised the company's share capital from three to 20 million Deutschmarks. At this point, the Porsche KG turnover was around 430 million Deutschmarks, generated by a workforce of 3,891 staff.

1971

In conjunction with the increase in capital completed the previous year, as of 1 April 1971, the company management of Dr. Ing. h.c. F. Porsche KG was expanded. Managing Director Ferry Porsche was now joined by Heinz Branitzki (Finance), Karlernst Kalkbrenner (HR), Ferdinand Piëch (Technical), F. A. Porsche (Design) and Michel Piëch (Administration) on the board. On 1 October, Ernst Fuhrmann took over the Technical department. An "External Development" division headed by Ferdinand Piëch was added, too.

The works made their first-ever start at the East African Safari Rally. Poland's Sobiesław Zasada and his co-driver Bień Marian took fifth place with a specially refurbished 911 S 2.4 (ST). Throughout the rally, they were accompanied by a plane that sent information and spare parts requirements ahead to the maintenance mechanics.

A 917/20 with a much larger body, which was a combination of short-tail and long-tail, was developed for the Le Mans race. The voluptuous curves inspired the Porsche design department to paint it pink, and to label it according to butcher-style cuts. The "Pink Pig" became the most photographed entrant at the 24-Hour classic but was forced to retire after an accident caused by a break defect.

At the East African Safari Porsche made use of support out of the aircraft.

Porsche 917/20 „Pink Pig" with distinctive design in Le Mans.

In summer 1971, the Development division comprising the Construction, Testing and Design departments gradually began to relocate from the main factory in Zuffenhausen to the new Weissach Development Centre (EZW). In addition to internal developments, the centre, also known as the "think tank", would go on to create – along numerous designs for Volkswagen – projects such as the compact four-cylinder engine for the Spanish car manufacturer Seat and a complete small car for the Soviet state-run company AvtoVAZ (Lada) in subsequent years.

The first few buildings forming part of the Weissach development centre taking shape.

To meet the strict exhaust gas regulations of the key sales market in the United States, Porsche developed the 2.4-litre six-cylinder engine, available since August 1971 in the models 911 T, E and S, which was capable of running on normal petrol.

The Porsche 911 S overhauled for model year 1972 was the worlds first series sports car to feature a front spoiler.

The 911 S was the first production sports car in the world to have a front spoiler.

As in the previous year, the 1971 racing season was dominated by the Porsche 917, and Porsche comfortably won the manufacturers' world championship with eight wins out of ten. The highlight of the season was another win at 24 Hours of Le Mans – this time by Gijs van Lennep and Dr Helmut Marko – in a Porsche 917 KH. With an average speed of 222.304 km/h and a distance covered of 5,335 kilometres, they set two records, the latter of which remained unbroken for 39 years. A 917 Langheck Coupé picked up another record: a top speed of 387 km/h was measured for the racing car on the Mulsanne Straight.

1972

At the beginning of the year, the partners of Dr. Ing. h.c. F. Porsche KG decided to turn the limited partnership (KG) into a limited company (AG) with share capital of DM 50 million at the beginning of the financial year on 1 August. Ferry Porsche became Chairman of the Supervisory Board, while Ernst Fuhrmann became spokesman for the Executive Board of Porsche AG. The Porsche and Piëch families' decision to cease appointing family members to operational management roles in future preceded the transition from limited partnership (KG) to limited company (AG).

Porsche Supervisory Board Chairman Ferry Porsche (left) and Executive Board spokesman Ernst Fuhrmann.

As part of involvement in the Canadian American Challenge Cup (CanAm), in June, the private Penske racing team entered the Porsche 917/10 Spyder with turbocharger for the first time in Mosport. Delivering up to 1,000 PS, the racing car dominated the racing series and won the CanAm championship for Porsche with victories in Road Atlanta, Mid-Ohio, Elkhart Lake, Laguna Seca and Riverside. In the European Interserie, the 917/10 didn't just win the fifth round at the Hockenheim Motodrom straight away – it took the whole championship.

George Follmer and Mark Donohue achieved a double victory for Porsche at the CanAm race in Laguna Seca.

In October 1972 Porsche presented a new basic car for racing and rally driving at the Paris Motor Show, the "Carrera RS" – the first Porsche 911 also known as the "Carrera". Initially launched as a homologation series of 500 units, the RS, with an appearance characterised by front and rear spoilers, was in such high demand from customers that Porsche produced 1,525 units of the high-performance sports car, which cost at least DM 34,000. Alongside the basic version, the new model was primarily sold in two versions, "Sport" and "Touring", and the racing version 911 Carrera RSR, of which 55 units were produced. The RSR made its first motorsport appearance at the Tour de Corse on 4 and 5 November 1972.

At the end of the year, the Porsche company employed a total of 3,950 employees. 14,265 cars in the 911 model series were sold by the end of the year. 59.6 per cent of these were the 911 T, 16.2 per cent the 911 E, 22 per cent the 911 S and 2.2 per cent the 911 Carrera RS 2.7.

The 911 Carrera RS 2.7 with the striking "ducktail" rear spoiler.

Porsche 911 T 2.4 Targa (US) and Porsche 911 E 2.4 Coupé.

1973

Following the previous year's transformation into a limited company, Dr. Ing. h.c. F. Porsche AG was officially entered in the Commercial Register on 1 March.

The most powerful racing car of its era, the Porsche 917/30 Spyder, made its racing debut in May. The superiority of the 882 kW (1,200 PS) circuit racing car was so evident that the rules of the CanAm series were changed at short notice to rule the 917 out of further participation in the 1974 season. In addition to the CanAm title picked up by the Porsche 917/30, a 917/10 won the European Interserie.

Porsche entered the manufacturers' world championships for prototypes with the 911 Carrera RSR. The works cars picked up an overall win (Müller/van Lennep) at the Targa Florio, taking place for the last time, and GT wins in the 6 Hours of Vallelunga, the 1,000 kilometres of Dijon and Spa. Customer teams also used the RSR in sports events and won races including the 24 Hours of Daytona and the European and German Rally Championships.

The Porsche 917/30 dominated the American CanAm series.

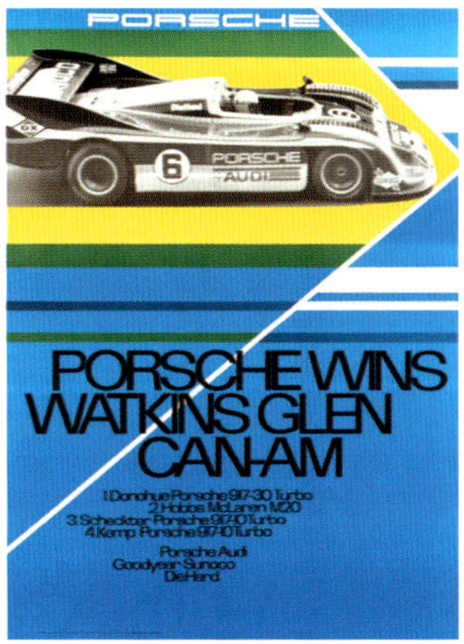

Herbert Müller and Gijs van Lennep triumphed at the 1973 Targa Florio in the Porsche 911 Carrera RSR.

The 911 Carrera RS 2.7 Coupé proved its sporting prowess with a successful record ten-kilometre run.

Huschke von Hanstein set two absolute world records and four international class records with a Porsche 911 Carrera RS 2.7 on the Volkswagen high-speed track at Ehra-Lessien. The former Porsche racing manager improved records over ten miles and ten kilometres from a standing start.

The 911 model series, now known internally as the G-series, was equipped with redesigned safety bumpers from September. These were intended to shape the appearance of the 911 until the start of the successor generation 964 in model year 1989.

From the 1974 model year, the G-series replaced the first 911 generation.

In addition, standard three-point seat belts and seats with integrated headrests ensured increased passenger protection.

At the Frankfurt IAA Motor Show, Porsche turned heads among car manufacturers worldwide with the "Long-Life Car Research Project" (FLA), a concept car designed for twenty years' use.

The "Long-Life Car Research Project" (FLA) was presented at the IAA in Frankfurt am Main in 1973.

1974

In the 1973/74 financial year, the remaining 50 per cent of the VW-Porsche-Vertriebsgesellschaft mbH was acquired by Porsche AG, making it the sole owner of the subsidiary.

The headquarters of VW-Porsche-Vertriebsgesellschaft mbH circa 1971 in Ludwigsburg.

The Porsche 911 Turbo, presented at the Paris Motor Show, was something of a sensation. The distinctive appearance of the car, known internally as the Type 930, includes its wider body, wide tyres, front spoiler and rear wing. The 911 Turbo, delivered since spring 1975, was the fastest series car of its era.

The 911 Turbo 3.0 was the first turbocharged series sports car from Porsche.

The immense potential of the new 911 Carrera RSR Turbo 2.1 manifested itself in June when it came second in the 24 Hours of Le Mans. The lessons learned in the Motorsport department during development of the racing car flowed directly into development of the production version of the 911 Turbo.

Herbert Müller behind the wheel of the 911 Carrera RSR Turbo 2.1 at the 24 Hours of Le Mans.

The significance of the Porsche 911 as a customer sports car was reflected by victories in numerous international championships. In 1974, these included the FIA GT World Cup, the European GT Championship and the IMSA GT and Trans-Am championships.

The Porsche works team's involvement in the Safari Rally in Kenya – probably the toughest rally competition of its time – ended with a second-place showing. In 1974, like two years before, just a small technical defect prevented a Porsche 911 Carrera RS 2.7 picking up the win.

The Brumos customer team won the American TransAM and IMSA GT series.

The Porsche 911 Carrera RS 2.7 "Safari" finished the East African Safari in second place.

On 20 September, work on the second section of the Weissach Development Centre was finally completed and testing facilities, workshops and laboratories were inaugurated. All development departments were now situated in a central location. This enabled optimal organisation of processing of customer orders from all over the world.

Aerial view of Porsche Werk 8 in Weissach, 1 July 1975.

1975

The „SAVE" project, designed to develop an innovative emergency vehicle, was initiated at the Weissach Development Centre in 1975. The system consisted of an emergency vehicle with a rescue unit and underwent development in various project phases until 1984.

The "SAVE" was created as part of a project to develop an innovative rescue vehicle.

Porsche AG was the first car manufacturer to use hot-dip galvanised sheet metal on both sides as standard. This allowed Porsche to increase the durability of corrosion-protected chassis for the long term and ensure a long-term guarantee against rust damage of up to six years.

Hot-dip galvanised body panels were used on the Porsche 911 from model year 1976 onwards.

Ludwigsburg-based VW-Porsche Vertriebsgesellschaft mbH was transferred to Dr. Ing. h.c. F. Porsche AG on 31 July for the start of the new financial year.

In August 1975, the American racing driver Mark Donohue set a new lap world record with a Porsche 917/30 Spyder in Talladega, USA. Its average speed was 355.86 km/h.

The Porsche 917/30 Spyder ahead of the world speed record in Talladega, USA.

Despite the lack of an official works team, an extensive customer sport sales portfolio resulted in an above-average track record at the end of the season. Alongside national title wins – such as the German Rally Championship – Porsche racing cars also won the Interserie and IMSA series.

Porsche customers dominated the list of winners at the IMSA racing series.

1976

Assembly of the Porsche 924 with a transaxle design, which was added to the sales range at the start of the year, was outsourced to Audi NSU Auto Union AG in Neckarsulm. In March 1972 Volkswagen had placed Development Order (EA) 425 for the design of a successor to the VW-Porsche 914. EA 425 had progressed to the preproduction stage when Volkswagen cancelled the project in late 1974 for reasons of product line policy. Porsche thereupon purchased the rights to the design. Due to huge global demand, production rose to 80 cars per day by July, and by the end of the financial year, the entry-level model accounted for a whopping 48 per cent of car sales. 150,684 of this entry-level Porsche had been built by the time its production run ended in 1988 with the 924 S.

The Porsche 924 was originally a development order for the Volkswagen Group.

In 1976, Porsche responded to the new Group 5 regulation in the manufacturers' world championship with the 935. Delivering 433 kW (590 PS), the 935, based on the Type 930, featured a fully lightweight construction. It won its first works forays in Mugello and Vallelunga and, after further triumphs, the world championship title made its way to Stuttgart at the end of the year.

In 1976, Porsche won both the manufacturers' world championship and the sports car world championship.

In addition to the manufacturers' world championship, a world championship for sports cars was announced for the first time in 1976. In light of this, the Porsche racing engineers developed another racing car, the 936, to be launched alongside the 935 customer sports car. Participation in Le Mans, which was already decided during the promising test drives, was rewarded with victory for Jacky Ickx and Gijs van Lennep. The sports car world championship crowned the turbocharged Porsche 936 with the maximum possible score. Having reached the summit, Porsche AG then stepped back almost entirely from racing, limiting itself to technical support for private teams.

On 6 November, the Supervisory Board of Dr. Ing. h.c. F. Porsche AG named the previous spokesman for the Board, Ernst Fuhrmann, CEO. Heinz Branitzki was appointed his deputy.

Porsche Chairman of the Executive Board Ernst Fuhrmann.

1977

Porsche presented a luxurious high-performance sports car to the public in spring in the shape of the 928. In addition to a V8 light-alloy engine, an aluminium chassis and the lane-correcting "Weissach rear axle", Porsche also broke new ground with the body shape of the 928, which was created under the direction of Head Designer Anatole Lapine.

The 928 with the transaxle construction was launched on the market from model year 1978.

Almost three decades after the production of the first Porsche sports car, 3 June saw the 250,000th car roll off the assembly line. This took place against the backdrop of extraordinarily good financial circumstances for the company, which saw the company end the 1976/77 financial year with a turnover of one billion Deutschmarks.

A speech by Ferry Porsche marking the production of the 250,000th Porsche.

The 911 model series was tightened up. Accordingly, from late summer 1977, only two derivatives of this perennial favourite were still on sale. The only remaining version with a naturally aspired engine was the 911 SC. Delivering 132 kW (180 PS), its appearance was inspired by the 911 Carrera. The portfolio's top model, the Porsche 911 Turbo at a cost of DM 78,500, underwent a performance boost. In addition to a new brake system with perforated brake discs, an intercooler boosted the performance of the engine, now boasting a 3.3-litre capacity, to 221 kW (300 PS).

The Porsche 911 Turbo 3.3 (front) and Porsche 911 SC from model year 1978.

As well as defending the manufacturer's world championship title with the 935, Porsche won the 24 Hours of Le Mans once again with the 936 (Jacky Ickx/Jürgen Barth/Hurley Haywood). Especially developed for the small division of the German Racing Championship with up to two litres of engine volume, the 935 "Baby" proved that Porsche wasn't only up to the task in large-capacity classes. At Hockenheimring, Jacky Ickx outclassed the competition by half a lap.

Reaching the finish line in the Porsche 936 for the fourth Le Mans victory for Zuffenhausen overall.

Commissioned by the Federal Ministry of Research and Technology, the firefighting system "O.R.B.I.T." was designed in the Weissach Development Centre in 1977. The purpose of the research project was to assist fire brigades with technical equipment fit for the future.

O.R.B.I.T. – a German acronym for optimised rescue and firefighting with integrated technical assistance.

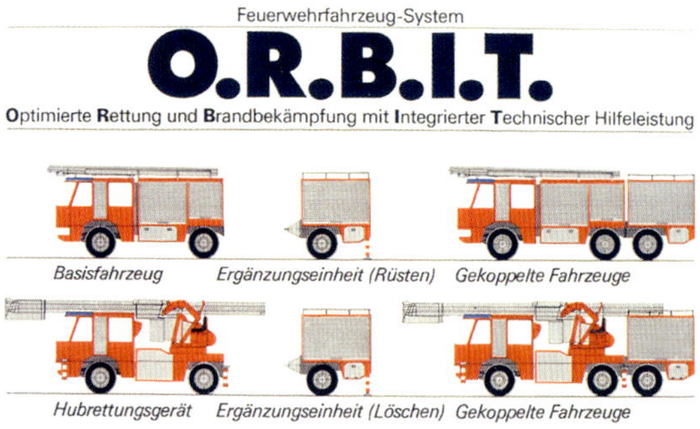

1978

The Porsche 928 was the first sports car to be voted "Car of the Year" by an international panel of journalists.

Prize-giving ceremony in Monaco, 1978.

Based on a wide-ranging portfolio of four-, six- and eight-cylinder sports cars, sales of DM 1.123 billion were recorded by the end of the 1977/78 financial year – an increase of 12 per cent. Due to the constantly growing car industry, this figure leapt an additional 20 per cent to DM 1.350 billion for the financial year 1978/79.

The Porsche 924 Turbo, visually recognisable from the cooling slots in the front and the 15-inch rims.

International racing victories in 1978 included the sixth win in the manufacturers' world championship, the European Sports Car Championship, the IMSA and Interserie and the European Hill Climb cham-

pionships. One special surprise was the Nicolas/Lavenne crew's triumph at the Rallye Monte-Carlo. The private drivers finished the rally as winners ahead of many works teams in a 911 Carrera.

A Porsche racing poster celebrating the triumphs of the 1978 season.

At the East African Safari Rally, Vic Preston Jr and Björn Waldegård started the 5,000-kilometre ordeal in two Porsche 911 SC cars. They finished in second and fourth places.

Group photo featuring the drivers, mechanics and engineers during the 1978 East African Safari Rally.

The 935/78 "Moby Dick" was the most powerful version of the Porsche 911 ever designed. However, the 621 kW (845 PS) racing car, equipped with two turbochargers, enjoyed a motorsports career spanning only four races. Following a win at the 6 Hours of Silverstone and an eighth overall placing in Le Mans, technical issues in Vallelunga prevented another victory.

The 935/78 "Moby Dick" was named after the whale from Herman Melville's novel on account of its shape.

The women's Porsche Tennis Grand Prix was held for the first time in Filderstadt in October. Tennis player Tracy Austin won the USD 35,000 prize money after defeating Betty Stove in the singles final.

The turn of the model year saw the arrival of a new addition to the sales portfolio: the Porsche 924 Turbo. The company answered many customers' wish for more power with this model, first delivered in early 1979. At 125 kW (170 PS), engine performance was 36 per cent higher than the naturally aspirated engine.

Tracy Austin triumphed at the first Porsche Tennis Grand Prix.

1979

Rudi Lins and Gerhard Plattner's 37,000-kilometre long-distance journey from New York to Vienna offered impressive proof of the reliability of the new Porsche 924 Turbo.

The 924 Turbo proved its reliability with a long-distance record drive.

Porsche appeared at the Frankfurt IAA booth with a surprise: inspired by the 924 Turbo, the 924 Carrera GT was a sports car equally suitable for both the road and the racetrack. The 924 Carrera GT was available to interested customers from the 1981 model year.

The Porsche 924 Carrera GT was a 154 kW (210 PS) and 150 kilogrammes lighter competition version.

The internal designation Type 995 referred to the prototype of a future sports car commissioned by the Federal Ministry of Research and Technology and created in 1979. The design of the four-seater prototype took particular consideration of consumption, safety and noise emissions.

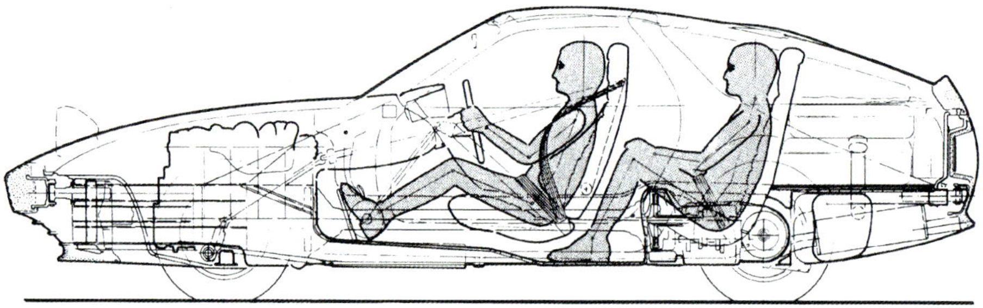

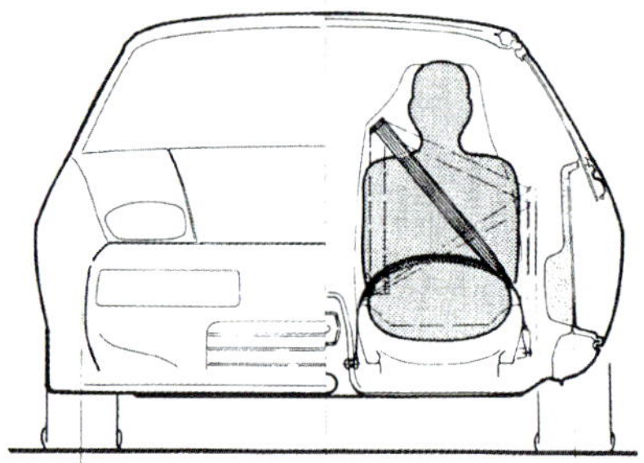

The Porsche 995 concept car was presented in 1979 as a four-seater sports car of the future.

Production of the Porsche 928 S began in autumn. Equipped with front and rear spoilers, the premium sports car delivered 221 kW (300 PS).

Ferry Porsche and the Porsche 928 S outside the Porsche Villa in Stuttgart.

For the North American market, Porsche offered engines designed for unleaded fuel with a regulated catalytic converter and lambda probe for the first time.

In light of reduced works activity, in 1979, it was primarily private teams who racked up 15 national and international championship wins for Porsche. These successes included triumphs in the German Racing Championship, the European Hill Climb Championship and the IMSA and TransAm championships. The towering achievement was another win in the manufacturers' world championship, achieved in no small part due to the success of private Type 935 cars in the 24 Hours of Daytona and Le Mans.

In a surprise announcement in December, Porsche declared its intention to take part in the Indianapolis 500 in 1980. However, issues with regulations prevented the Porsche-Interscope "Indy" from racing.

Customer sports teams racked up a series of successes for Porsche, including world championship titles.

Danny Ongais driving the Porsche-Interscope "Indy", boasting over 900 PS.

From model year 1989, the 911 was also available with all-wheel drive. The first 200 demonstration cars were collected from the skid pad by the Porsche dealer organisations in Weissach.

1980 — RACES AND RALLIES — 1989

"The all-wheel drive for the Paris–Dakar Rally was a highlight. That came right after the triple win in Le Mans in 1982."

Peter Falk

1980

Heralding a new Porsche generation, the 924 drew public attention as it swung into action in the motorsport season. When the Porsche 924 Carrera GT made its debut at the 24 Hours of Le Mans, all three works cars made the finish line. The transaxle sports car racked up 25 wins at the American SCCA Championship.

The Porsche 924 was specially converted for endurance racing in Le Mans.

With the racing cars 935 and 911 SC, Porsche won three European championship titles and the World Endurance Championship, as well as the IMSA and TransAm championships. Another highlight of the season was taking the top three spots in the 24 Hours of Daytona.

A special victory: Porsche took the top three spots at the 24 Hours in Daytona.

Cockpit of the Airbus A310 developed by Porsche.

In collaboration with the company Airbus Industries, the engineers at Weissach began designing an ergonomic cockpit layout for Airbus passenger aircraft. Work on the development order, with the Porsche project designation Type 2603, took place from 1980 to 1984.

Porsche expanded the guarantee for cars from model year 1981 to seven years' long-term protection against rusting. The guarantee covers the entire body, which is now made entirely of hot-dip galvanised sheet steel.

Porsche granted a seven-year guarantee against rusting through on the hot-dip galvanised body.

At the eighth "International Conference on Experimental Safety Vehicles" in October, Porsche AG presented the research prototype Type 960, known as "P.E.S.". Akin in form to the Porsche 928, the entire front of the test car was made from aluminium.

Porsche Type 960, 1980.

1981

German-American Peter W. Schutz took over the role of CEO of Dr. Ing. h.c. F. Porsche AG from Ernst Fuhrmann on 1 January.

Peter W. Schutz took the role of Chairman of the Executive Board.

A Porsche dual-clutch transmission (PDK) developed in Weissach enabled the shifting of gears without any interruption in traction. The first test run of the transmission took place in 1983 in a Porsche 956.

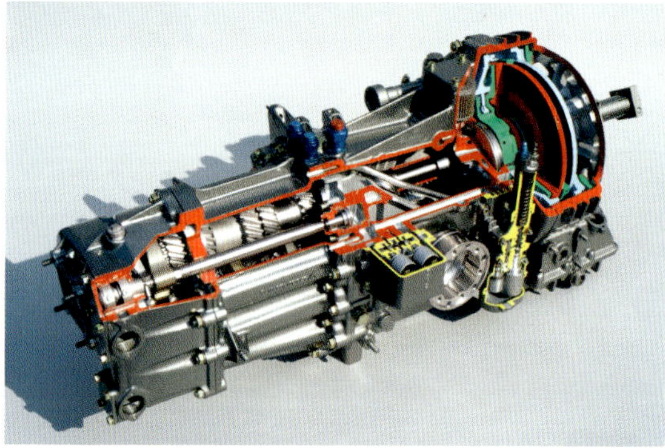

The Porsche dual-clutch transmission (PDK) was first tested in racing.

Porsche entered a prototype of what was to become the 944, known as the 924 GTP Le Mans, in the 24 Hours of Le Mans. The new type of car proved extremely reliable, allowing Walter Röhr and Jürgen Barth to finish the race with a class win. Another Le Mans entrant from Stuttgart marked the company's 50th anniversary with an overall win: the Porsche 936/81 Spyder with the driver team Jacky Ickx and

Derek Bell was created shortly before the race, based on a 936 from the 1976 racing season, and equipped with a newly developed 2.65-litre six-cylinder engine.

In 1981, the Porsche 936 won the 24 Hours of Le Mans for the third time after 1976 and 1977.

Porsche presented the 944 series version for the first time at the Frankfurt IAA Motor Show. With a 120 kW (163 PS) four-cylinder engine, the 944 was positioned between the entry-level 924 and the 911 SC. An all-wheel-drive cabriolet prototype based on the Porsche 911 Turbo and showcased at the same time drew special attention.

Porsche 911 Turbo 3.3 4x4 Cabriolet at the IAA in 1981.

The boxer engine from the 911 model series was used to create the Porsche aircraft engine PFM 3200, designed for small aircraft. Low fuel consumption, minimal noise emissions and easy operation set the engine apart.

The six-cylinder boxer engine: functional in the air as well as on the road.

In the period from 1981 to 1984, the department for foreign developments for external companies at Weissach Development Centre worked on designing a new model range of Linde forklifts, based on the latest research into ergonomics. It marked the beginning of a long-standing cooperation.

Third-party orders from external companies were also processed at the Weissach development centre.

1982

The market launch of the Porsche 944 at the beginning of the year attracted huge attention from the press, trading partners and customers worldwide. Before production ceased in 1991, a total of 163,302 of the four-cylinder sports cars in the 944 model series were made.

The 944 was the second four-cylinder transaxle model from Porsche, alongside the 924.

The 911 SC Cabriolet, presented at the Geneva Motor Show in March, was a milestone in Porsche model history. The open-top sports car featured a newly designed three-bow cabriolet top. Sheet steel profiles integrated into the fabric roof not only ensure dimensional stability at high speeds, but also a low noise level and good rollover protection.

After 19 years, Porsche launched another cabriolet onto the market.

The 1982 motorsport year was all about new records. Accordingly, as well as finishing in the top five places in Le Mans, the Porsche brand succeeded in clinching almost every class win and special category. The top three spots were taken by the new 956, the first Porsche racing car with a monocoque chassis and ground effect. Group 5 was won by a Porsche 935 K3, Group 4 by a 934, the IGTX category by a 935 and the IGT category by a 924 Carrera GTR. In addition to triple victory in Daytona, other highlights of the season included the manufacturers' and drivers' championships in the World Endurance Championship and the IMSA racing series.

The new 956 scored a triple victory on its first appearance at Le Mans.

The Measuring Centre for Environmental Technology (MZU), the world's first autonomous emissions testing facility, opened in late autumn 1982. The roller dynamometers located in Weissach created computer-controlled simulations of real drivingresistances, thus making it possible to reproduce various test cycles. In a climate chamber, test series can be carried out under a wide variety of climatic and altitude conditions.

Testing the 928 S at the Measuring Centre for Environmental Technology (MZU) in Weissach.

A new high-bay warehouse with fully automatic storage and retrieval, connected to the assembly line by driverless transport systems in the underground channel, was put into operation in Zuffenhausen.

1983

Porsche presented a very special high-performance car at the Frankfurt IAA Motor Show: the "Group B" prototype. Designed for endurance racing, the prototype's body featured soft lines and curves that would reappear in later generations of the 911. Named the Porsche 959, the version of the "Group B" prototype ready for production caused a stir worldwide from 1985 on.

The 959 prototype already hinted at the super sports car's subsequent appearance.

Voted "Motorsports car of the year", the Porsche 956 was also available to private racing teams in 1983. This meant the racing car's series of wins was not limited to works cars. This allowed the previous year's success at the 24 Hours of Le Mans to be surpassed. In the end, Porsche took places one to eight. At the end of the season, the many individual victories were consolidated into international titles such as the manufacturers' and endurance world championships and the IMSA manufacturers' and drivers' championships.

The racing poster marking the Le Mans victory in 1983.

With 6:11 minutes, Stefan Bellof achieved an incredible record on the north loop of Nürburgring.

In training for the 1,000 kilometres at Nürburgring, Stefan Bellof reached an average speed of over 200 km/h in a Porsche 956 for the first time in the racetrack's history.

Commissioned by British racing team McLaren International and financed by Saudi businessman Mansour Ojjeh, the "TAG Turbo" engine made its debut in a McLaren MP4 Formula 1 racing car in summer 1983. Designed as a 1.5-litre six-cylinder V-engine with turbocharging under the guidance of Hans Mezger, the light and compact drive delivered up to 1,000 PS.

With a total of 25 Grand Prix victories and three world championship titles, the "TAG Turbo made by Porsche" became the dominant engine in Formula 1 from 1984 to 1986.

The Porsche TAG Turbo engine dominated Formula 1 from 1984 to 1986.

In late summer, on the 20th birthday of the Porsche 911, the 911 SC was replaced by the models known as 911 Carrera. The overhauled 3.2-litre boxer engine with digital engine electronics now delivered 170 kW (231 PS). The Coupé, Targa and Cabriolet versions were still available and joined by the Turbo-look version available for the 911 Coupé for model year 1984.

From model year 1984, the basic 911 models were known as Carrera.

1984

The Porsche 911 Carrera 3.2 4x4 (953) took a surprise win at the Paris–Dakar Rally. The all-wheel-drive prototype sporting the colours of the sponsor Rothmans was driven by René Metge and Dominique Lemoyne. This marked the first win for a sports car in the desert marathon.

Internally, the 911 Carrera 3.2 4x4 Paris-Dakar was also known as 953.

A special homologation series of 20 units of the Porsche 911 SC RS was delivered to private customers with motorsporting ambitions in early 1984.

Only 20 of the 911 SC RS were built for homologation purposes.

In spring, Porsche increased the limited company's share capital from DM 50 million to DM 70 million and at the same time introduced 50 per cent of the share capital as non-voting preferred shares on the stock exchange. The issue price of the securities traded since 4 May was previously set at DM 780.

Under the control of Porsche Enterprises Inc., Porsche Cars North America Inc. took over the import of Porsche sports cars to North America on 1 September. The agreement in place with the Volkswagen subsidiary Volkswagen of America Inc. since 1970 lapsed on 31 August. Other newly established entities included Porsche España S.A. and Porsche Italia S.p.A.

Porsche Cars North America vehicle transporter in the USA.

As in the previous two years, the 956 was the dominant endurance racing car and secured the manufacturers' world championships for Porsche once again. Although no works team started at Le Mans, seven private Porsche 956 cars picked up the top spots.

The TAG Turbo engine began its series of triumphs in the McLaren Formula 1 racing car MP4/2. With 12 out of 16 races won, Niki Lauda and Alain Prost took the two top spots of the drivers' world championship uncontested.

McLaren was the undisputed winner of the Formula 1 season.

Ferry Porsche, 1984.

Baden-Württemberg's Minister-President Lothar Späth awarded Ferry Porsche the title of Professor to mark his 75th birthday.

At the IAA in Frankfurt, the Porsche 959 highlighted the Stuttgart-based sports car manufacturer's technical expertise. From the biturbo boxer engine with water-cooled four-valve cylinder heads to the electronically controlled chassis and all-wheel-drive system through to the aerodynamically optimised body, Porsche showed what was possible in the automotive industry with the 959.

1985

With 162 kW (220 PS) and a cd value of 0.33, the driving performance of the Porsche 944 Turbo, introduced in January, was among the best in its class. It was the world's first car to be able to deliver the same performance with or without a catalytic converter.

The 944 Turbo shot from 0 to 100 km/h in 6.3 seconds and achieved peak speeds of 245 km/h.

Only 292 Porsche 959 super sports cars were made.

The anti-lock braking system (ABS), available as a special feature in the Porsche 928 since 1983, was installed as standard from late summer 1985. The Porsche 928 S was also available with a regulated three-way catalytic converter. At the same time, the rust guarantee for all Porsche sports cars from the 1986 model year was increased to ten years.

The Porsche 928 S from model year 1986.

Equipped with the drive brakes and chassis components of the Porsche 944, the 924 S represented the final development stage of the Porsche entry-level model from September.

The Porsche 924 S from model year 1986 onwards.

In October, for the first time, the Porsche 959 no longer started with the naturally aspirated engine of the 911 Carrera 3.2, but with the since-completed 2.8-litre biturbo. Its first assignment was the nine-day Pharaohs Rally in Egypt, where Saeed Al Hajri and John Spiller took first place right away.

Porsche 959 at the Pharaos Rally.

The Porsche 962 brought the manufacturers' world championship home to Zuffenhausen for the fourth time running. In addition, Hans-Joachim Stuck secured the title of endurance driving world champion together with Derek Bell during his first Porsche season. In 1985, private Porsche teams achieved great successes in Monza and Le Mans, as well as in the American IMSA series and in the German Sports Car Championship. In Formula 1, Alain Prost won the world championship in a McLaren with a TAG Turbo engine.

In 1985, Porsche honoured its racing successes with a winners' poster.

1986

At the start of the year, Porsche Italia S.p.A. in Padua began its Porsche import activities. As of 1986, Porsche AG also held all shares in the British import company Porsche Cars Great Britain Ltd.

French drivers René Metge and Dominique Lemoyne pipped Jacky Ickx and Claude Brasseur to victory in the 14,000-kilometre Paris–Dakar Rally in another 959. As a "flying service car", with project manager Roland Kussmaul at the wheel and his colleague Wolf-Hendrik Unger as co-driver, the third 959 with starting number 187 took sixth place.

Porsche Aviation Products Inc. in Washington, Illinois, was established as a sales company for Porsche aircraft engines. The Porsche aircraft engine PFM 3200 had proven its reliability to the public the previous year during an around-the-world flight.

The Porsche wind tunnel in the Weissach development centre.

With the measurement centre for aerodynamic research and testing, the Weissach Development Centre had one of the most effective wind tunnels in international automotive engineering since 1986. In addition, an ultra-modern paint facility with eco-friendly technology ensured optimum paint application in Zuffenhausen since spring.

Available from August 1986 the Porsche 944 S was advertised with the attributes of "environmentally-friendly technology" and "sporty functionality". The transaxle series 928 also received a comprehensive upgrade after nine years of construction and presented the Porsche 928 S4 with a new eight-cylinder engine and aerodynamically enhanced body.

Alongside the sports car world championships for teams and drivers, in 1986, Porsche racked up wins at the Sport Auto Super Cup, the IMSA and Interserie and in the Formula 1 World Championship with the

Porsche 928 S4.

TAG Turbo engine. The 24 Hours of Le Mans ended with the eleventh Porsche overall win in total. The circuit version of the Porsche 959, the Type 961, achieved an IMSA/GTX class win.

The super sports car 959 demonstrated its superiority off the tarmac, too.

An eco-friendly racing series, the Porsche 944 Turbo Cup, was set up in 1986. The identical cars were equipped with a regulated three-way catalytic converter as standard. The overall winner of the first edition was Joachim Winkelhock.

Porsche used the 944 Turbo Cup to gain experience with catalytic converters in full-load operation.

1987

The Weissach Development Centre was expanded to include a modern crash test facility with permanently installed steering and surveillance technology. It was used to test cars from other manufacturers as commissioned by customers, as well as Porsche sports cars.

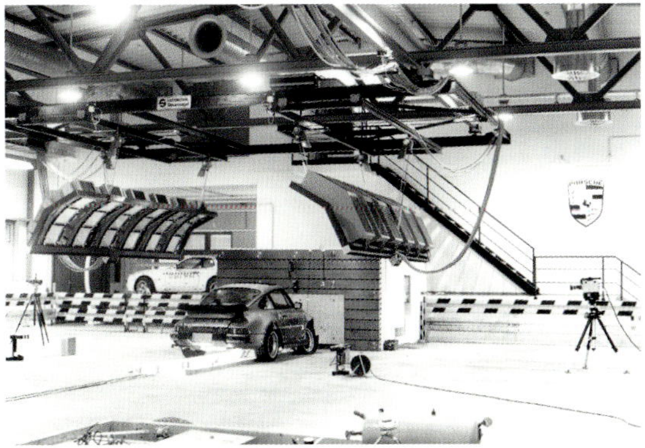

Crash test facility at the Porsche Development Centre in Weissach.

The Porsche 911 Carrera Speedster drew huge crowds at the Frankfurt IAA International Motor Show. The open-top one-seater continued the philosophy of the 1950s Porsche Speedster and was fully designed for pure driving pleasure with a spartan array of accessories.

From model year 1989, the 911 Carrera Speedster was available in a limited edition of 171 cars.

Spring 1987 saw the addition of the Porsche 911 Turbo Targa and the 911 Turbo Cabriolet to the 911 model series. They were joined by the Porsche 911 Carrera Clubsport, around 100 kilogrammes lighter than the series version of the 911 Carrera.

The Porsche 911 Turbo Cabriolet (right) and 911 Turbo Targa from model year 1987.

The 944 Turbo S special model, with the 250 PS engine from the Turbo Cup.

The model range of four-cylinder sports cars saw the addition of the Porsche 944 Turbo S in late summer. Initially limited to a special edition of 1,000, the 184 kW (250 PS) unit of the sports coupe also went on to be used in the regular Porsche 944 Turbo from model year 1989.

The announcement of the withdrawal from the World Endurance Championship heralded a change in the commitment of Porsche to motorsport. A decision was made to participate in the American CART series in 1988 and work began on developing a single-seater racing car. The Le Mans works participation of the 962 C, which set a new brand record with the twelfth overall victory for Porsche at the Sarthe, was a particular success.

Brand record for Porsche in Le Mans.

Porsche CEO Peter W. Schutz departed the company on 31 December. The Supervisory Board appointed previous Deputy CEO Heinz Branitzki to replace him.

Heinz Branitzki.

1988

Porsche made a foray into unfamiliar motorsport terrain with the Type 2708 CART. Developed for the American CART series, this racing car was a brand-new design with an aluminium and plastic monocoque. A V8 racing engine with 551 (750 PS), running on methanol, powered the car. Porsche treated the 1988 CART championship races primarily as test runs.

The new Porsche body opened in Zuffenhausen on 25 August. The bodies for the 911 and 928 models were now assembled on the same production line – the capacity of the facility amounted to 120 bodies per day.

The Indianapolis 500: the season highlight of the American CART series.

The Porsche bodyshop in Zuffenhausen was inaugurated in 1988.

Porsche upgraded its range of models for model year 1989 with an array of improvements. The worldwide innovation of the tyre pressure monitoring system, or "RDK", made its way into the Porsche 928 S4. Porsche had previously successfully trialled the technology over years in racing. The Porsche 911 Turbo got a new five-gear transmission with hydraulic clutch actuation.

The last model year of the 911 Turbo 3.3 in the G-series.

The press launch of the Porsche 944 S2 took place in October 1988. Its three-litre four-valve engine was not only the world's biggest, most powerful four-cylinder naturally aspirated engine. Thanks to the harmonious way its power developed, it also attracted particular praise from journalists.

The new three-litre engine for the 944 S2 from model year 1989 onwards.

911 Carrera 4, model year 1989.

Marking the 25th anniversary of the 911, the all-wheel-drive Porsche 911 Carrera 4, presented in November, heralded the arrival of a new generation of the hit model. Highlights of the sports car known internally as the 964 included the new 3.6-litre engine and an aerodynamically enhanced body with plastic front and rear sections and automatically extending spoiler.

1989

At the beginning of the year, the open-top Porsche 944 S2 Cabriolet was assembled by the American Sunroof Corporation (ASC) in Weinsberg near Heilbronn. A Porsche 944 Turbo Cabriolet was added to the range in 1991 as an especially exclusive four-cylinder version.

In the model year 1991, only 528 of the 944 Turbo Cabriolet were made.

On 1 April, Dutchman Harm Lagaaij replaced Anatole Lapine, head of the design department Porsche Style at Porsche since 1969, who departed at the end of the previous year. At the Frankfurt IAA Motor Show, the Weissach-based designer showcased the Roadster prototype "Panamericana", created to mark the 80th birthday of Ferry Porsche.

Ferry Porsche (centre) with his sons Hans-Peter, Gerhard, Ferdinand Alexander and Wolfgang (from left to right).

From April onwards, Porsche Leasing GmbH (PLG), a wholly owned subsidiary of Porsche AG, managed a growing proportion of Porsche used- and new-car customers.

The Porsche 911 from generation 964 was produced in three bodywork versions from model year 1990.

In July 1989, Porsche called time on production of the 911 Carrera and the 911 Turbo in the "G-series", marking the end of an era. The successor model was the Porsche 911 Carrera 2 from Generation 964, equipped with a regulated catalytic converter and anti-lock braking system. Over the course of model year 1990, this car and the 911 Carrera 4 were both available in Coupé, Targa and Cabriolet versions.

From November onwards, Porsche began offering a new transmission known as Tiptronic, enabling automatic and manual switching of gears.

The Tiptronic gearbox in the Porsche 911: comfort and dynamism combined.

Another new arrival: the Porsche 928 GT, with 243 kW (330 PS) and a maximum speed of 275 km/h, represented a new peak of performance in the Porsche sales portfolio.

At the Mid Ohio race, Italian Teo Fabi won the first CART victory for Porsche in a Quaker State Porsche with a March chassis. Porsche placed among the top four in nine out of 15 races in total.

Poster for the race that marked the first Porsche victory in the American CART series.

Porsche 928 GT in front of the Nürburg castle.

The silhouette of the Porsche 911 Carrera 3.4 Coupé of the 996 generation was glued in its original size in the design studio.

1990 — TURN-AROUND — 1999

"That's only possible with a motivated workforce who know how tough it will be, but that it's worth fighting for."

Wendelin Wiedeking

1990

Ferry Porsche (left) with his son Ferdinand Alexander Porsche.

In the Supervisory Board meeting that followed the Porsche general meeting in March, Ferry Porsche was named Honorary Chairman of the Supervisory Board. New additions to the board were Michel Piëch and Ferdinand Alexander Porsche, who followed in his father's footsteps as Chairman of the Supervisory Board. Arno Bohn became Chairman of the Executive Board of Porsche AG on 10 March; his predecessor Heinz Branitzki took retirement.

Equipped with the wider body and large rear spoiler, as per tradition, the 911 Turbo heralded the arrival of the third generation of the turbocharged 911 at the 60th edition of the Geneva Motor Show.

From 0 to 100 km/h in five seconds and top speeds of 270 km/h: the 911 Turbo.

In spring, work began on assembling the Mercedes-Benz 500 E, a production order for Mercedes-Benz AG. The order from Untertürkheim comprised extensive development work and the construction of the bodyshell, plus final assembly. By 1995, the former Porsche body shop in Zuffenhausen had produced 10,479 of the zippy Mercedes saloon.

Final inspection of the Mercedes-Benz 500 E, assembled by Porsche.

Olaf Manthey won the opening race of the Porsche Carrera Cup.

The Porsche Carrera Cup celebrated its world premiere in Zolder, Belgium, on 1 April and replaced the Porsche 944 Turbo Cup racing series. The winner of the opening race contested by 40 identical Porsche 911 Carrera 2 Cup race cars was touring car specialist Olaf Manthey.

Together with British specialists March, the racing engineers at Weissach developed a racing car chassis with a carbon-fibre monocoque, tailor-made for the Porsche Indy engine, for the 1990 CART season. However, regulatory changes at short notice reduced any chances of victory for the Porsche-March 90 P to nil.

The Porsche-March 90 P was thwarted by last-minute rule changes.

At the Birmingham Auto Show, the Porsche 911 Carrera RS from Generation 964 sparked enthusiasm among car lovers. The weight-optimised, higher-performance Carrera RS served as a homologation series for the N/GT group.

Porsche 911 Carrera RS.

1991

Porsche AG expanded its shareholdings in January to include Porsche Financial Management Services Ltd., based in Dublin, Ireland. In April, it was followed by the company K.K. Porsche Engineering Japan.

In early 1991, Porsche AG reorganised the sale of its cars in Germany. The number of dealerships was cut by a quarter to 80 Porsche centres, responsible for sales and customer service for the sports car from Zuffenhausen. From August, Porsche centres offered a guarantee for Porsche used cars that was unique worldwide.

Porsche became the first car manufacturer to equip all its models with driver and passenger airbags as standard from summer onwards. The introduction of eco-friendly water-based paint was another innovation.

As a successor to the Porsche 944, the Porsche 968 was available in Coupé and Cabriolet versions from late summer 1991. Until production ceased in 1995, a total of 11,245 of the four-cylinder transaxle sports car Porsche 968 were produced.

Crash-testing a Porsche 911 Carrera S Cabriolet with driver and passenger airbags.

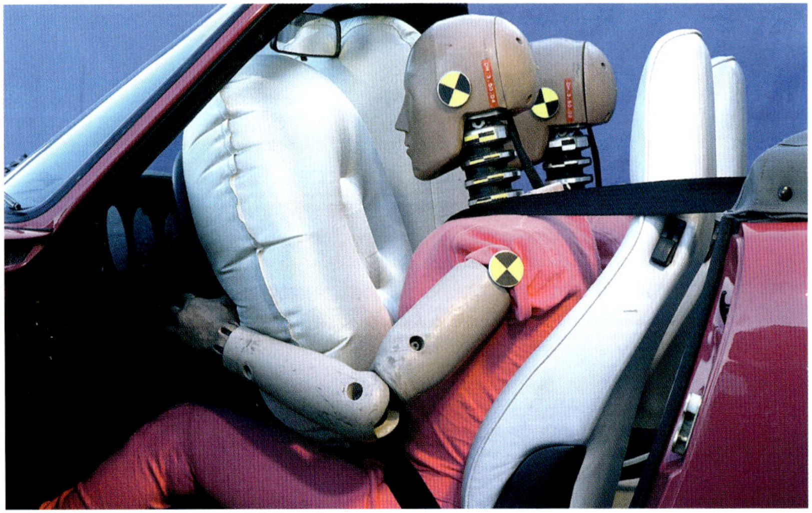

The 968 entered the market as the final model in the Porsche transaxle era.

The Porsche 928 GTS was presented at the IAA as the technical apex of the 928 model series, which had now existed for 14 years. Extensive comfort features and the overhauled 5.4-litre eight-cylinder engine made the Porsche 928 GTS the ideal endurance racing car.

Porsche 928 GTS, model year 1992.

The 3.5-litre twelve-cylinder Grand Prix engine developed by Porsche was exclusively available to the Footwork Arrows racing team in the 1991 Formula 1 racing season. However, an unfavourable start to the project and financial difficulties saw the partnership come to a premature conclusion at the season's end.

The Footwork Porsche FA 12 at the Grand Prix in Montreal.

1992

Porsche AG tackled falling sales figures due to the difficult global economic situation with an extensive package of measures. Under the leadership of Dr Wendelin Wiedeking, Director of Production and Materials Management since October 1991, new organisation and production workflows were introduced and hierarchy and process levels underwent root-and-branch reform as part of the concept of "lean management". The "Kaizen" philosophy introduced by a Japanese delegation of production experts saw the creation of the Porsche Improvement Process (PIP), designed to boost the company's performance and productivity sustainably, with involvement from all employees. As part of a stripped-down approach to work and processes, rationalisation measures such as short-time working and the loss of a total of 1,850 jobs were required.

Porsche production in 1992.

Wendelin Wiedeking.

Chairman of the Executive Board Arno Bohn departed the company on 30 September. He was replaced by Dr Wendelin Wiedeking, who became spokesman for the Executive Board on 1 October.

At the same time as internal campaigns to increase the company's competitive edge, moves were made to develop the model range for the long term. The exclusive special series of the Porsche 911 Turbo S, in production since spring, was followed by three more automotive innovations from Porsche at the Paris Motor Show: as an uncompromising Clubsport version of the four-cylinder series, the Porsche 968 CS attracted attention with a low sales price and amazing performance. As an open-top individual car, the Porsche 911 Carrera 2 Speedster from the Generation 964 expanded the range, while the newly designed Porsche 911 Turbo 3.6 took a top position in the sales portfolio alongside the Turbo S.

The 911 Carrera 2 Speedster: only 930 units were made.

1993

At the Detroit Motor Show, Porsche Executive Board spokesman Wendelin Wiedeking and Porsche Style Chief Harm Lagaaij presented the future-oriented Boxster design study, whose design recalled the 550 Spyder and 718 racing cars.

The Boxster design study enjoyed a highly positive reception.

As a basis for the ADAC GT-Cup, the Motorsport department in Weissach began work on a special series of the road-legal Porsche 968 Turbo S and racing-ready Turbo RS from January 1993. The four-cylinder turbo engine of the RS delivered up to 257 kW (350 PS) and higher racing speeds of 280 km/h.

Alongside a version for the ADAC GT-Cup, the 968 Turbo RS was also available in an endurance version for Le Mans.

By resolution of the general meeting, the share capital of Porsche AG was increased by DM 17.5 million to DM 87.5 million in spring 1991. To that end, existing shareholders were offered ordinary and preference shares in a subscription ratio of 4:1 at a price of DM 575 each. The company received a total of DM 201 million from the newly issued shares.

Ferry Porsche and son Ferdinand Alexander at the "30 years of 911" anniversary.

As part of a joint venture, Audi and Porsche developed a high-performance sports estate car known as the Audi Avant RS2. The Avant RS2 was manufactured alongside the Mercedes-Benz E 500 in the Zuffenhausen production facilities from October.

With its five-cylinder in-line engine, the Audi Avant RS2 delivered 232 kW (315 PS).

At the start of the new financial year 1993/94, on 1 August, the Supervisory Board named Dr Wendelin Wiedeking Chairman of the Executive Board of Porsche AG.

A new step in the evolution of the Porsche 911 made its worldwide debut at the Frankfurt IAA Motor Show in September. The Porsche 911 Carrera from Generation 993 featured a redesigned aluminium chassis, a harmonious design and a 3.6-litre boxer engine with many upgraded details.

A debut for the fourth generation of the 911.

1994

The Porsche 911 Carrera Cabriolet from Generation 993 enjoyed its world premiere at the Detroit Motor Show in January. The open-top sports car made its European debut shortly afterwards in March at the Geneva Motor Show.

A self-erecting wind deflector was available for the first time in 1994 for the Porsche 911 Carrera Cabriolet.

The automatic Tiptronic S transmission with paddle shifters on the steering wheel of the Porsche 911 Carrera created a new Porsche driving experience from model year 1995.

Tiptronic S with paddle shifters on the steering wheel.

The Porsche customer development team began developing the Opel Zafira compact van on behalf of Adam Opel AG based on the specified concept. The Opel Zafira made its debut at the 1998 Paris Motor Show.

Porsche customer development: Opel Zafira.

The Porsche 962 made an unexpected comeback at 24 Hours of Le Mans. Modified as part of a rule change for the GT1 class, three of the 962 Dauer Le Mans GT sports cars were built in Weissach in cooperation with Jochen Dauer Racing. Alongside a double victory in the GT1 and triple triumph in the GT2 class, the 24 Hours of Le Mans came to a close with the 13th overall win for Porsche.

At Le Mans in 1994, Porsche picked up the overall win as well as top spots in the GT1 and GT2 class.

In August, Porsche Consulting GmbH was founded as a wholly owned subsidiary of Porsche AG. This allowed the sports car manufacturer to make its experience from its own restructuring process available to other companies as a consulting service.

The Porsche 911 Carrera 4 was available in Coupé and Cabriolet versions from September for fans of all-wheel-drive cars. Thanks to the visco clutch, the four-wheel-drive sports car guaranteed optimum traction in all driving situations.

The chassis of the all-wheel-drive Porsche 911 Carrera 4.

In November, Porsche AG took part in an international presentation with 19 other manufacturers at the Beijing World Trade Center. The family car on show, C 88, was a Porsche prototype designed especially for the Chinese market. Wendelin Wiedeking gave his presentation speech in Mandarin, an event that merited a page of coverage in the German magazine *Der Spiegel*.

The C88 was a Porsche prototype designed especially for the Chinese market.

From late autumn 1994, Porsche AG sold the extensive accessories line "Selection". The product range featured everything from textile collections and leather accessories to a series of luggage specially designed for the Porsche 911.

1995

From spring, Porsche AG offered an expanded selection of accessories for personalising Porsche sports cars under the brand name Tequipment.

A new generation of the Porsche 911 Turbo was presented in Geneva in March. The highlight of the all-wheel super sports car was the biturbo engine equipped with the OBD II exhaust gas control system, considered to be the world's lowest-emission standard car engine. The hollow-spoke aluminium rims, used for the first time in series car manufacturing, were another innovation.

The top speed of the biturbo engine was 300 km/h.

The Porsche 911 Carrera RS offered a purely sporty driving experience, with an excellent power-to-weight ratio thanks to 221 kW (300 PS) and 1,270 kilogrammes. This was only surpassed by the Porsche 911 GT2, accelerated by a 316 kW (430 PS) biturbo engine. The road version with racing ambitions was the sportiest and fastest model in the range.

Porsche 911 GT2, model year 1995.

After three economically difficult years and accumulated losses of DM 450 million, Chairman of the Executive Board Dr Wendelin Wiedeking was able to establish a turnaround in fortunes at the end of the 1994/95 financial year. The company broke even as a result of the improved sales situation and measures to improve productivity in particular.

Assembling the Porsche 911 Carrera at Zuffenhausen.

After the four-cylinder 968 series and the eight-cylinder 928 series were discontinued at the end of the 1995 model year, Porsche offered further attractive model variants for the classic sports car with the 911 Carrera 4S and 911 Targa models presented in September. Another innovation in the Targa was the electric glass roof retracting behind the rear window.

Porsche launched the Targa in model year 1967.

At the IAA in September, Porsche presented a wide range of international travel experiences for the first time under the Porsche Travel Club name.

As of 1 October, Porsche Leasing GmbH changed its name to Porsche Financial Services GmbH. It continued to offer its customers tailored leasing and financing agreements.

1996

Located in the Weissach Development Centre, the Car Industry Emissions Centre (ADA) was jointly established in January 1996 by the German car manufacturers Audi, BMW, Mercedes-Benz, Volkswagen and Porsche.

The GT racing car Porsche 911 GT1 made its debut in June at the 24 Hours of Le Mans. After a double victory in the GT1 category and securing second and third place in the overall rankings, the TWR-Porsche WSC Spyder driven by Team Joest also achieved the overall win.

The 1,000,000th Porsche rolled off the assembly line in Zuffenhausen on 15 July. The Baden-Württemberg motorway police received the anniversary car, a Porsche 911 Carrera Coupé.

Porsche racing cars took the three top spots at the 24 Hours of Le Mans in 1996.

The production milestone of the 1,000,000th Porsche sports car was celebrated on 15 July 1996.

The title of "Factory of the Year", awarded by Produktion weekly and business consultants A.T. Kearney, honoured Porsche AG as Germany's most efficient production facilities.

Engine production at the main Porsche plant in Zuffenhausen.

Presented in mid September, the Porsche Boxster set new benchmarks in sports car construction. The long-awaited mid-engined Roadster was enthusiastically received by the media, dealers, importers and customers alike. Over 10,000 orders were placed worldwide before customers had even laid eyes on the new two-seater. Another new addition to the model range was the Porsche 911 Carrera S with its wide Turbo body.

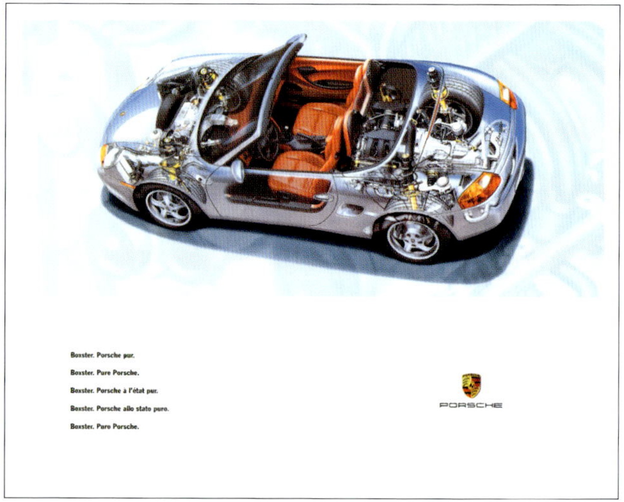

The first-generation Porsche Boxster, known internally as 986.

Porsche Engineering Services GmbH (PES GmbH), with headquarters in Bietigheim-Bissingen, was founded in October. The new subsidiary's team of around 120 engineers was mainly intended to assist with work on customer orders at the Weissach Development Centre.

1997

For the first time in its history, Porsche AG entered the bond market. The Euro bond worth DM 200 million was used to finance future growth, including a Porsche off-road vehicle as part of the expansion of the business base.

The Porsche works team entered the highlight of the motorsport season, the 24 Hours of Le Mans, with the 1997 version of the 911 GT1. Ultimately, however, it was a Joest TWR-Porsche WSC Spyder prepared according to World Sports Car rules that brought home the 15th overall Porsche win to Zuffenhausen. A private Porsche 911 GT2 also picked up a class win. Technical problems shortly before the race's end saw both Porsche 911 GT1 cars drop out. A small series of 20 Porsche 911 GT1 road-legal cars was built for sports-loving customers.

Porsche 911 GT1 '97.

Porsche AG entered into a joint venture with the Harley-Davidson Motor Company in summer 1997. The companies agreed on joint engine production in Milwaukee, Wisconsin, in the US. This represented the provisional pinnacle of the collaboration between Porsche and the motorcycle manufacturer, which had existed since 1969. The highlight of the partnership was the "Revolution" engine type developed by Porsche and ready for series production, used in the Harley-Davidson "V-Rod" since 2002.

Decades of collaboration between Porsche and Harley-Davidson culminated in engine production for the V-Rod.

Porsche tackled demand for the Boxster, which exceeded all expectations, by expanding production capacity. In addition to production in Zuffenhausen, from September, work began on manufacturing the Porsche Boxster in Uusikaupunki, Finland, at Valmet Automotive Inc. 109,213 units were produced in Finland until the model change of the series known internally as 986 in 2004.

The Valmet Automotive Inc. plant in Uusikaupunki, Finland.

A year after the Porsche Boxster was successfully launched on the market, the new 911 Carrera was presented at the Frankfurt IAA Motor Show in September with the slogan "Evolution 911". As a completely new development, the new sports car generation was powered for the first time by a water-cooled four-valve six-cylinder boxer engine. The design interprets the classic lines of the 911 in an innovative way and boasts an impressively low cd value of 0.30.

1998

In line with the international strategy for Porsche to manage all sales in its key export markets itself, the subsidiary Porsche Japan took over imports for the Japanese market in January.

At the Porsche AG general meeting, Porsche CEO Dr Wendelin Wiedeking was able to announce for the 1997/1998 financial year not just an increase in earnings from DM 165.3 million to DM 324.4 million, but also the highest annual surplus in the company's history, at DM 276.9 million. Shareholders received a dividend payment amounting to DM 42.9 million.

The Porsche type range for the 1998 model year.

Professor Ferry Porsche passed away aged 88 on 27 March in Zell am See. He was laid to rest in the burial chapel on the Schüttgut family estate on 30 March. The official memorial service took place on 3 April at the Zuffenhausen Porsche plant.

Ferdinand Anton Ernst "Ferry" Porsche.

Porsche announced the decision to build a third series on 3 June. In addition to the Porsche 911 and the Boxster, the company aimed to tap into new sales markets with an off-road multipurpose vehicle.

Porsche Cayenne design sketch.

At the French endurance classic Le Mans, the Porsche 911 GT1 with its carbon-fibre chassis outdid tough competition and brought home a double victory and the 16th overall Porsche win to Zuffenhausen. Laurent Aïello, Allan McNish and Stéphane Ortelli were at the wheel of the winning car, Jörg Müller, Uwe Alzen and Bob Wollek took second place.

A double triumph at Le Mans marked the 50th anniversary of Porsche sports cars in 1998.

Early summer saw the debut of the Porsche 911 Carrera Cabriolet, presented at the Geneva Motor Show. This was the world's first open-top car with side airbags suitable for a cabriolet. With the 3.4 litre boxer-engine it achieved a top speed of up to 280 km/h.

Porsche 911 Carrera Cabriolet, model year 1998.

As another stage in the product strategy, the Porsche 911 Carrera 4 with permanent all-wheel drive was presented in Paris in autumn. In addition to the Coupé and Cabriolet versions, customers now had the choice between a manual six-gear transmission and the five-programme Tiptronic S.

Showcased at the Paris Motor Show, the Porsche 911 Carrera 4 was a new addition to the range from October 1998.

1999

The Porsche 911 GT3 turned heads at the Geneva Motor Show.

From 1 January onwards, the subsidiary Porsche France, based in Cergy-Pontoise near Paris, organised the sales of Porsche sports cars in the world's fifth-biggest car market.

The Porsche 911 GT3 was presented at the Geneva Motor Show as a successor to the Porsche 911 Carrera RS. Designed as a sporty niche model with dry sump lubrication and a separate oil tank, the GT3 was also available as a Clubsport version for racing purposes. The Porsche 911 GT3 R earned its first triumphs straight away at the 24 Hours of Le Mans, where it achieved a class win and the 13th overall ranking.

A new pinnacle in the Zuffenhausen product strategy: the 911 Turbo, presented at the Frankfurt IAA Motor Show. As the first Porsche Turbo generation, the top-range model was also available with five-gear Tiptronic S, bringing an end to the Turbo-free era for those customers who were used to driving automatic. At the same time, Porsche introduced a global innovation with the Porsche Ceramic Composite Brake (PCCB). The brake set new benchmarks for responsiveness, fade stability, weight and lifespan.

With its six-cylinder boxer engine with biturbo charging, the 911 Turbo broke the 300 km/h mark.

The update to the Boxster Generation 986 made in August at the turn of the model year included the basic model, now boosted to 2.7-litre capacity, as well as the 185 kW (252 PS) Boxster S, powered by a 3.2-litre unit.

In Germany, Porsche opened up new sales channels and founded Porsche Deutschland GmbH. The subsidiary Porsche Middle East managed clients in 13 different countries.

In November, Porsche AG joined the "German Economy Foundation Initiative" for the compensation of forced labourers. The foreign workers who were forcibly employed by Porsche KG under the Nazi regime received compensation. Seven individuals affected were each compensated with a one-time payment of DM 10,000.

The Porsche Boxster, available in two performance classes from model year 2000.

The new Porsche Museum at the company headquarters in Stuttgart-Zuffenhausen opened on 31 January 2009.

2000 — MODEL DIVERSITY — 2009

"The Porsche legend lives and will never die."

Wolfgang Porsche

2000

Porsche boss Wendelin Wiedeking was able to announce record sales and earnings figures for the financial year 1998/1999 at the general meeting. Sales had risen by 25.5 per cent to DM 6.18 billion. At DM 698.2 million, the group result had more than doubled. With a profit margin of 11.3 per cent, Porsche was one of the world's most profitable car manufacturers at the beginning of the new millennium. For the first time in company history, its own equity exceeded investment capital.

The special series Porsche 911 Carrera 4 "Millennium" was created to mark the turn of the millennium. Only 911 of the anniversary model with its premium interior and Turbo-look rims with a high-gloss finish were made.

The limited-edition model 911 Carrera 4 "Millennium", only available in the special colour Violet Chromaflair.

Together, Saxony's Minister-President Kurt Biedenkopf, the Mayor of Leipzig Wolfgang Tiefensee and Porsche CEO Wendelin Wiedeking broke ground on construction of the new Porsche factory in Leipzig on 7 February. Series production of a five-seater "Sports Utility Vehicle" was planned for 2002 with the quality seal "Made in Germany".

Breaking ground on the new Porsche factory in Leipzig.

The production-based prototype Porsche Carrera GT made its global debut at the Paris Motor Show in the Louvre on 28 September. The high-performance sports car's carbon-fibre chassis and the ten-cylinder naturally aspirated engine were based on pure racing technology.

The press unveiling of the Porsche Carrera GT took place in September 2000 at the Paris Motor Show.

In the American Le Mans Series (ALMS), works drivers Bob Wollek, Jörg Müller, Lucas Luhr and Sascha Maassen won the GT ranking title in the Porsche 911 GT3 R from Dick Barbour's team.

Porsche customer teams achieved success in the GT classes with the 911 GT3 R.

With the Porsche Centre Stuttgart, 19 October saw the first Porsche factory branch open in Zuffenhausen. It is located in the immediate vicinity of production. The self-contained, rounded facade became a model for other Porsche centres worldwide.

The Porsche Zentrum Stuttgart on Porscheplatz.

2001

On 8 January, the Porsche 911 GT2, presented at the Detroit Motor Show, became the new peak performer in the sales portfolio. With 340 kW (462 PS) and 620 Newton metres, the GT2 was the most powerful series sports car yet produced by Porsche. The super sports car was equipped with ceramic brake discs as standard.

The 911 GT2: the pinnacle of performance among Porsche production sports cars.

Despite significant development costs for the third Cayenne series, Porsche CEO Dr Wendelin Wiedeking was once again able to announce a huge rise in sales, revenue and production to shareholders at the general meeting on 19 January. Pre-tax profits of DM 848.5 million amounted to a profit margin of 11.9 per cent.

From 22 February, Porsche sports cars intended for export were transported by rail to the international Port of Emden. The new loading station in Kornwestheim was inaugurated in the presence of Federal Chancellor Gerhard Schröder and railway boss Hartmut Mehdorn.

Opening ceremony for the rail loading station in Kornwestheim on 22 February 2001.

A stock split decided at the Porsche general meeting in January was performed on 23 April. The conversion from par value shares to no-par value shares took place at a ratio of 1:10.

Porsche Engineering Group GmbH (PEG), a wholly owned subsidiary of Porsche AG, officially began operating on 1 August. PEG's purpose was the global management and marketing of customer developments.

The new 3.6-litre engine of the updated Porsche 911 Carrera boasted a variable valve lift and performance of 235 kW (320 PS) from September 2001.

The all-wheel-drive 911 Carrera 4S and the 911 Targa were new additions to the sales portfolio.

Porsche AG tapped into a new market with an official sales office in the capital of China, Beijing. Alongside Hong Kong, the "Porsche Centre Beijing" was the brand's second foothold in the promising Chinese market.

911 Carrera Coupé, model year 2002.

2002

Porsche was once again able to put forward record figures at the annual general shareholders' meeting. Wendelin Wiedeking announced turnover of 4.4 billion euros for the financial year 2000/2001 – an increase of 21.8 per cent. The untaxed group result increased by 36.5 per cent to 592.4 million euros, while the profit margin increased to 13.3 per cent. For the first time in company history, over 50,000 cars were sold in one financial year.

From February, Singapore-based Porsche Asia Pacific took charge of the concerns of Porsche importers in the region, from South Korea to New Zealand. Together with the regional offices Porsche Middle East and Porsche Latin America, a total of 37 importers were being looked after. In addition to another 25 independent importers, the entire global Porsche network now spanned 550 retailers.

From 7 to 30 June, the updated versions of the Porsche Boxster and Boxster S of Generation 986 were presented to the international media in Grottaferrata, near Rome. The Boxster's 2.7-litre engine now delivered 168 kW (228 PS), while the 3.2-litre unit on the Boxster S boasted 191 kW (260 PS). The new VarioCam technology, now also in use in the Boxster engine, boosted performance and reduced consumption and emissions.

The Boxster and Boxster S from model year 2003 onwards.

The guests at the official inauguration of Porsche Leipzig GmbH on 20 August were treated to a surprise presentation of the Porsche Cayenne. The multipurpose vehicle was presented by Chancellor Gerhard Schröder, who tightened the last of the first car's screws, and Porsche CEO Dr Wendelin Wiedeking, ahead of its official world premiere.

Gerhard Schröder tightening the final screw on the first Cayenne at the Porsche Leipzig factory.

At the Paris Motor Show, the sporty all-terrain Porsche Cayenne multipurpose vehicle was officially presented to a global audience. The car, designed in cooperation with Volkswagen under the development leadership of Porsche, was initially available in Cayenne S and Cayenne Turbo versions.

The Cayenne S and Cayenne Turbo outside the Leipzig customer centre.

2003

Contrary to all economic trends, the Porsche Executive Board was able to announce pre-tax earnings up 40 per cent to 828.9 million euros in January. Shareholders benefited from another record result in the form of a special dividend worth 14 euros per share. At the end of the financial year 2001/2002, the Porsche share had risen in value by 1,739 per cent within ten years, from 26.74 to 465 euros.

The Porsche Carrera GT enjoyed its world premiere in Geneva on 3 March 2003. Its design was as extraordinary as its driving performance: the Carrera GT accelerated from zero to 200 km/h in just 9.9 seconds, reaching top speeds of 330 km/h. The limited-edition high-performance sports car was built at the Porsche plant in Leipzig – the 450 kW (612 PS) V10 naturally aspirated engine was supplied by the Zuffenhausen engine plant.

A new generation of the Porsche 911 GT3 went on sale in dealerships in March. The engine performance of the extreme sports car with the dominant rear wing rose to 280 kW (381 PS) due to an increase in engine speed.

With the construction of a new service centre in Bietigheim, the three subsidiaries Porsche

The Carrera GT hit the road in racing style.

The 911 GT3: a new addition to the model range.

The new service centre in Bietigheim.

Deutschland, Porsche Financial Services and Porsche Consulting were brought together under one roof from mid 2003.

Porsche continued its product strategy with the 2004 model year range, which was available since late summer: the 40th anniversary of the Porsche 911 saw the presentation of the special model "40 Years of 911", limited to 1,963 units. For the first time since 1987, a Porsche 911 Turbo was available as a cabriolet once again. The road-legal Porsche 911 GT3 RS was available as the basis for motorsport events. The Zuffenhausen-based company expanded the range of sporty multipurpose cars with the six-cylinder Cayenne. Another exclusive version of open-top driving arrived in dealerships in October in the form of the Porsche 911 Carrera 4S Cabriolet.

Porsche 911 Carrera "40 Years of 911"

2004

A throwback to the legendary racing car, the special Boxster S model "50 Years of 550 Spyder", presented at the Detroit Motor Show, boasted enhanced engine performance and extensive optional equipment. The special edition was limited to 1,953 cars.

Porsche celebrated the anniversary of its first motorsports car built as a racing car with a special edition model.

In the sports car category, readers of the magazine auto motor und sport voted the Porsche Carrera GT the outright winner in the "Best Cars" competition.

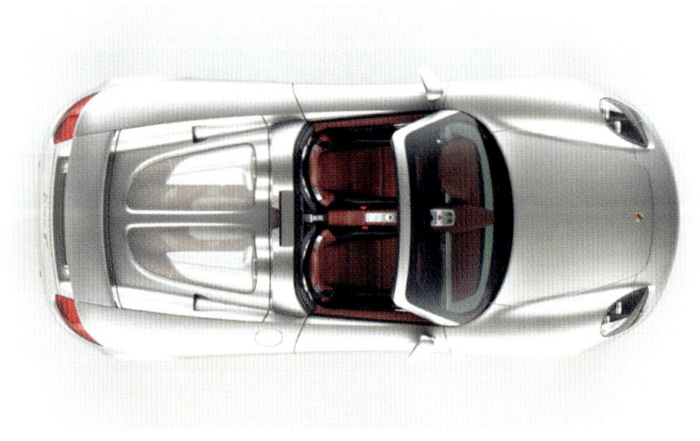

By May 2006, a total of 1,270 Porsche Carrera GT cars had been produced.

After a 12.6 per cent rise in pre-tax earnings to 933 million euros in the 2002/2003 financial year, there were signs of continued growth in the new financial year. At the general meeting, which was set to be held for the first time in the new Leipzig Trade Fair, CEO Wendelin Wiedeking predicted annual sales of more than 75,000 cars for the first time.

In 2004, the Porsche annual general meeting was held at the Leipzig Trade Fair.

In the upper market segment, Porsche was expanding the product range to include the 911 Turbo S Coupé and 911 Turbo S Cabriolet. The new "S" versions were set apart by the increased performance potential of the engine and the standard ceramic brake system PCCB.

The 911 Turbo was available as a coupe and cabriolet in model year 2005.

The market launch of the latest generation of the Porsche 911 was set to take place simultaneously on 16 July in 85 German Porsche centres. The models, known internally as the 997, were initially available in 911 Carrera and 911 Carrera S versions.

The 911 Carrera S Coupé and 911 Carrera Coupé from model year 2005.

In addition to the Porsche 911 Carrera S, the new Porsche Boxster and the Boxster S of Generation 987 were presented to an international audience at the Paris Motor Show.

The new Boxster with 2.7-litre capacity in Paris.

2005

From January, Porsche produced the V8 engines for the Cayenne in the newly built engine plant in Zuffenhausen. In addition to series engine production, engine prototypes were also manufactured in the 9,000-square-metre facilities.

At the Detroit Motor Show, the convertible version of the new Porsche 911 heralded a second stage in the version strategy for what was now the sixth generation of the 911. Like the Coupé, the 911 Cabriolet was available as a Carrera and Carrera S. In autumn, the 911 Carrera 4 and 911 Carrera 4S joined the Porsche model family. Both were also available in Coupé and Cabriolet versions.

The all-wheel-drive Porsche 911 Carrera 4S Cabriolet offered extraordinary driving stability and traction.

At the general meeting on 28 January, Ferdinand Alexander Porsche was appointed Honorary Chairman of the Supervisory Board. His son Oliver Porsche succeeded him as a member of the Supervisory Board. For the eleventh time in a row, CEO Wendelin

Wiedeking was able to announce new record sales (up 15 per cent) and turnover (up 6.8 per cent) for the past financial year. Pre-tax profits rose to 1.238 billion euros.

Ferdinand Alexander Porsche and son Ferdinand Oliver.

One of the highlights of the Frankfurt IAA Motor Show was the Porsche Cayman S, a sporty, ambitious mid-engined sports car positioned between the Boxster and the 911.

The Cayman S: a closed mid-engined sports car based on the Porsche Boxster S.

In August, the Supervisory Board and Executive Board decided to develop and produce a fourth Porsche series. The four-seater coupe known as the Panamera would go on the market in 2009.

Panamera design sketch, 2005.

The media and investments banks alike were equally surprised by news of the Porsche stake in Volkswagen AG. With the acquisition of around 20 per cent of the voting capital, Porsche became the largest single shareholder in the Wolfsburg-based automotive group. The strategic goal of the stake was to safeguard future plans with development partner Volkswagen in the long term and simultaneously to defend against a hostile takeover of Volkswagen AG by financial investors. Wendelin Wiedeking and Head of Finance Holger P. Härter went on to represent Porsche interests on the Volkswagen AG Supervisory Board.

From autumn, Porsche took part in the American Le Mans Series (ALMS) with the new sports prototype RS Spyder. On its first outing at Laguna Seca in California, the RS Spyder achieved pole position, a lap record and class victory.

Porsche RS Spyder.

2006

In January, Porsche AG issued a dollar hybrid bond with a volume of one billion US dollars and a Euro bond with a total volume of two billion euros. Both investments were so oversubscribed by investors that the demand ultimately amounted to more than ten billion euros.

At the Porsche general meeting on Killesberg in Stuttgart, the share price, sales and profits once again satisfied the Porsche shareholders. For the previous financial year 2004/05, group earnings before taxes increased by 8.9 per cent to 1.238 billion euros. CEO Wendelin Wiedeking was also able to announce new record values for the half-year figures for the 2005/06 financial year: compared to the same period of the previous year, sales increased by 15 per cent to 3.25 billion euros.

At the Geneva Motor Show, Porsche AG presented the new 911 Turbo on 28 February. The sixth generation of the top model in the 911 series was the first serially produced car with a petrol engine to have a turbocharger with variable turbine geometry (VTG). At the same time as the 911 Turbo, the new 911 GT3 was also presented to global audiences in Geneva. Its 3.6-litre naturally aspirated engine developed an output of 305 kW (415 PS),

The Porsche 911 emerged with new technology.

corresponding to a specific performance per litre of 115.3 PS. The GT3 thus set a new benchmark for road-legal, serially produced sports cars with naturally aspirated engines in this capacity class.

In early April, the Porsche Cayenne Turbo S celebrated its German trade fair premiere alongside the 911 Turbo and the 911 GT3. At the Auto Mobil International (AMI) exhibition centre just a few kilometres from the Porsche Factory Leipzig, the sporty off-road vehicle impressed in no small part due to its 383 kW (521 PS) V8 drive.

The Porsche Executive Board decided to produce the four-door sports coupe Panamera in the Leipzig plant on 19 May. In addition to a significant increase in production capacity, around 600 new jobs were set to be created at Porsche Leipzig GmbH. Around 400 jobs would be created in Zuffenhausen and Weissach by 2009.

With 383 kW (521 PS), the Cayenne Turbo S offered peak performance among Porsche SUV models.

Alongside the Cayenne, the Panamera was also produced at the Porsche Leipzig factory.

For the first time ever, an LMP2 racing car picked up an overall win in the ALMS.

At the third round of the American Le Mans Series (ALMS) in Mid Ohio, the Porsche RS Spyder fielded by the American Penske team on 21 May achieved a double victory. This was the first time that a racing car in the LMP2 class picked up an overall win in the ALMS.

Stuttgart saw the inauguration of a new multipurpose event hall in the form of the Porsche-Arena on 27 May. The Porsche Tennis Grand Prix took place for the very first time at the new Porsche-Arena from 30 September to 8 October.

Entry to Porsche-Arena.

The entry-level model Cayman was presented alongside the more powerful Cayman S.

Introduced in autumn 2005, the Porsche Cayman S saw the addition of a new version in late July. The Cayman was powered by a six-cylinder boxer engine with a 2.7-litre capacity and performance of 180 kW (245 PS).

The new 911 GT3 RS was launched on the European market in October. This version of the GT3, reduced in weight to 1,375 kilogrammes, served as a base and homologation model for use in various racing series.

The Porsche 911 GT3 RS: a remarkable link between a road-going sports car and racing car.

The Porsche RS Spyder finished its first full LMP2 season by topping the manufacturer, team and driver rankings. The sports car prototypes, also available for private customer teams, were produced in the new motorsports centre in Weissach, which opened in July.

In the sixth generation of the 911, all-wheel drive was a feature of all Targa derivatives.

With the 911 Targa 4 and the 911 Targa 4S, Porsche presented two new versions of the 911 series in November. The most striking features of the two all-wheel-drive sports cars were the large glass roof and the rear window suitable for opening.

A change of leadership took place in the Sales, Marketing and HR departments at the turn of the year. Klaus Berning succeeded Hans Riedel as Head of Sales on 1 November. In May 2007, Thomas Edig, who had been with the company since 1 December, took over from Harro Harmel as Head of HR and Employment.

Four years after the model's introduction, Porsche AG presented the second generation of its sports off-road vehicle, the Cayenne, in early December. The new models Cayenne, Cayenne S and Cayenne Turbo feature more powerful lower-consumption engines with direct fuel injection. The active chassis Porsche Active Suspension Management (PASM) was suitable for expansion to include the new roll stabilisation system Porsche Dynamic Chassis Control (PDCC), used for the first time by Porsche.

The second generation of the sporty off-road Cayenne offered more performance with less consumption.

At the company accounts press conference on 6 December, CEO Wendelin Wiedeking presented another record year of revenue for Porsche AG. Group earnings for the previous financial year 2005/2006

before tax rose to 2.110 (previous year: 1.238) billion euros. The Supervisory Board proposed to the general meeting on 26 January 2007 that the dividend be increased to 5.94 euros per ordinary share and six euros per preference share. In addition, to compensate for the high one-time income, plans were made to issue a special dividend of three euros per ordinary and preference share.

2007

At the turn of the year 2006/2007, the building shell of the new Porsche Museum was mostly completed. The actual exhibition area was built as a steel framework, requiring a total of 6,000 tons of steel. The 24,000-square-metre museum building, located in the immediate vicinity of the main plant, was scheduled for completion in the second half of 2008.

The Porsche Museum taking shape on Porscheplatz in Zuffenhausen.

At 4.44 p.m. on 5 January, an investor paid more than 1,000 euros for a Porsche preference share for the first time.

In the first six months of the financial year 2006/2007, Porsche achieved a pre-tax result of 1.59 billion euros (previous year: 277.8 million euros). In addition to the improved model mix in the car business, Chairman of the Executive Board Dr Wendelin Wiedeking also attributed the well-above-average leap in earnings to special factors relating to the involvement with Volkswagen.

Dr Wolfgang Porsche.

The Porsche AG Supervisory Board appointed Dr Wolfgang Porsche as its new chairman on 26 January. He succeeded Professor Helmut Sihler, who chaired the Supervisory Board for 14 years. Hans-Peter Porsche was a new addition to the supervisory body. After 27 years on the Supervisory Board, Dr Walther Zügel also resigned his post; he was succeeded by Professor Ulrich Lehner.

The Porsche Supervisory Board authorised the Executive Board in an extraordinary meeting on 24 March to increase the stake in Volkswagen AG from 27.3 per cent to up to 31 per cent of ordinary shares and thus to trigger a mandatory offer. In addition, there were plans to set up a holding structure with the primary aim of separating business operations from investment management. The intention was to spin off the business operations of Porsche AG into a wholly owned subsidiary. The parent company, which would then act as a holding company, would also be converted into a European stock corporation, Societas Europaea (SE).

In June 2007, the subsidiary Porsche Switzerland officially took over the import and sale of Porsche cars, spare parts and accessories from AMAG Automobil- und Motoren AG.

The 100,000th Porsche 911 of Generation 997 rolling off the production line.

Barely three years after the start of production, on 11 July, the 100,000th car in the 911 Generation 997 rolled off the assembly line at the main plant in Zuffenhausen.

At the Frankfurt IAA Motor Show in September, Porsche turned heads with three new innovations. The Cayenne GTS was 24 mm lower and delivered 298 kW (405 PS). A new generation of the 911 Turbo inspired performance-hungry open-air enthusiasts

The innovative new Porsche 911 Turbo Cabriolet and Porsche Cayenne GTS meet the mid-engined sports cars Cayman S and Boxster.

with 353 kW (480 PS) and 620 Nm torque. The most powerful series 911 of all time, the GT2 delivered an impressive 390 kW (530 PS).

The Transsyberia Rally, which started on 3 August, ran from Moscow to Mongolia. Porsche launched a small series of 25 Cayenne S Transsyberia cars for the event. After 7,108 kilometres, victors Rod Millen and Richard Kelsey reached the finish line in Ulaanbaatar.

The 25 Cayenne S for the Transsyberia Rally in the Porsche Leipzig factory.

In its second season in the ALMS, the Porsche RS Spyder continued to set the pace. The LMP2 racing car defended its title and racked up no less than eight overall and eleven class wins in 2007.

Resounding victory for the RS Spyder at the LMP2 championship at the American Le Mans Series.

In early December, Porsche showcased the special Boxster RS 60 Spyder model at the Bologna Motor Show. Its colour, shape, features and philosophy recall the 1960s era of Porsche motorsports.

The special edition Boxster S "RS 60 Spyder" was limited to 1,960 cars.

2008

With the 911 GT3 Cup S and the 911 GT3 RSR, Porsche Motorsport expanded its range for customer teams with two additional types of racing car. The Cup version complied with FIA GT3 Championship rules. The endurance specialist was built entirely in Weissach. The RSR rolled off the assembly line in Zuffenhausen, alongside the series 911 cars.

The 911 GT3 Cup S, designed exclusively for racing.

On 1 February, after just five years in production, the 200,000th Cayenne rolled off the assembly line in Leipzig. It was a Cayenne Turbo.

Production at the Porsche Leipzig factory.

After a construction period of less than a year, the first section of the building for the new central spare parts warehouse in Sachsenheim went into operation on 15 April.

Central spare parts warehouse in Sachsenheim.

April marked the world premiere of the new Cayenne Turbo S at the 2008 Auto China trade fair. The 404 kW (550 PS) model underlined the major significance of the Asian market for Porsche.

The Lava Grey Metallic paint colour was reserved for the Cayenne Turbo S.

The traditional 24 Hours of Nürburgring ended on 25 May with an overall victory for Porsche for the third time in a row.

In the 2008 season, the RS Spyder also took part in the European Le Mans Series (ELMS). After class wins in Barcelona, Monza and Spa, the RS Spyder sports cars fielded by customer teams achieved a double victory in the LMP2 class at Le Mans.

Porsche achieved a hat-trick at the prestigious Nürburgring 24 Hours.

The RS Spyder scored victories at the European Le Mans Series, too.

June saw trade journalists presented with the updated Generation 997 of the 911 Carrera and Carrera S models, equipped for the first time with direct fuel injection (DFI) and the Porsche dual-clutch transmission (PDK) as an optional extra. This new transmission boasted seven forward gears and replaced the previous Tiptronic.

The overhauled 997 generation featured different design and technology from its predecessor.

Late July saw the presentation of the all-wheel-drive versions 911 Carrera 4 and Carrera 4S. With the update, the electronically controlled Porsche Traction Management (PTM) replaces the previous all-wheel drive with viscous multi-plate clutch.

The sixth-generation 911 product range was expanded to include all-wheel-drive models.

The fastest transition between generations to date in the 911 family continued as early as September with the introduction of the new 911 Targa 4 and 4S.

The 911 Targa in model year 2009 featured an innovative glass roof.

2009

On 28 January, the new Porsche Museum was inaugurated and opened to the public for the first time three days later.

The opening Porsche Museum exhibition.

In the second half of January, the new Generation 987 Boxster was presented to the national and international press in Sicily. The focal points were the newly designed six-cylinder boxer engines with 2.9 litres in the Boxster and 3.4 litres in the Boxster S and the Porsche dual-clutch transmission (PDK), now available for the Boxster for the first time.

The second generation of the Boxster, known internally as 987, was launched on the market.

At the Geneva Motor Show, the new 911 GT3 celebrated its world premiere on 3 March. Due to an increase in capacity to 3.8 litres, the proven six-cylinder naturally aspirated engine now delivered 320 kW (435 PS).

The new 911 GT3 was even more powerful and dynamic than its predecessor.

Porsche launched a production car with a diesel drive for the first time.

The second new Porsche presented in Geneva was the Cayenne Diesel. The smooth-running V6 engine produced a performance of 176 kW (240 PS) from a capacity of three litres. It also boasted a remarkable torque curve, reaching the peak of 550 Nm as early as 2,000 revolutions per minute.

On 6 March, the 250,000th Cayenne rolled off the assembly line in Leipzig. The motoring milestone was a white Cayenne Diesel with a six-cylinder V-engine, delivered to a customer in Austria.

The Cayenne topped sales charts in 2009.

The new Porsche Panamera was presented to journalists on the 94th floor of the Shanghai World Financial Center on 19 April. It then made its first

The Porsche Panamera S on the 94th floor of the World Financial Center in Shanghai.

major public appearance at Auto Shanghai, China's leading motor show.

In July, the Supervisory Board of Porsche SE decided to begin preparations for a capital increase of five billion euros, the entry of an investor and the continuation of negotiations on the basic agreement with

Dr Wolfgang Porsche at Werk 2 on 23 July 2009

Volkswagen AG. The aim was to create an integrated automotive group. A partial step along this path would be Volkswagen AG's stake in Porsche AG. By stepping down, Wendelin Wiedeking made way for this new direction. Dr Wolfgang Porsche appeared deeply moved as he addressed the workforce in the pouring rain.

Following Dr Wendelin Wiedeking's departure, the Supervisory Board named Michael Macht Chairman of the Executive Board of Porsche AG on 23 July.

Michael Macht became Chairman of the Executive Board on 23 July 2009.

At Porsche Exclusive, the special model 911 Sport Classic was created based on the 911 Carrera S from Generation 997. Its global debut took place in September at the Frankfurt IAA Motor Show. The fixed rear spoiler, nicknamed "ducktail" in reference to the

911 Carrera RS from 1972, was particularly striking. In addition, the double-dome roof, the widened rear of the 911 Carrera 4S, a special front lip spoiler and 19-inch wheels in the Fuchs rim design gave the 911 Sport Classic a unique look. Within 48 hours of going on sale, all 250 cars were snapped up for 201,682 euros apiece.

The Porsche 911 Sport Classic was manufactured at Porsche Exclusive.

From 9 to 25 October, around 600 journalists from Germany and worldwide discovered the new 911 Turbo in Lisbon. The crowning glory is the new engine with 3.8-litre capacity, direct fuel injection and a turbocharger with variable turbine geometry (VTG).

The 911 Turbo was also available as a coupe and cabriolet in model year 2010.

Only 1,944 Boxster Spyder cars were made.

At the Los Angeles Auto Show in early December, the new Boxster Spyder celebrated its world premiere. The 3.4-litre six-cylinder engine with direct fuel injection delivered 235 kW (320 PS).

Just three months after sales began, the 10,000th edition of the new Panamera was produced on 14 December at the Leipzig plant. The 368 kW (500 PS) Panamera Turbo in platinum silver metallic was delivered to a customer in Singapore.

Making the Panamera at the at Porsche Leipzig.

With the Taycan Porsche launched its first full-electric sports car.

2010 — BRAND IDENTITY — 2019

“Our sports car of the future will definitely be recognisable as a Porsche.”

Michael Mauer

2010

After just six-and-a-half months of construction, the topping out ceremony for the new paint shop in Zuffenhausen took place on 1 February.

A new paint shop for the main Porsche plant in Zuffenhausen.

With the slogan "Porsche Intelligent Performance", Porsche showcased three hybrid world premieres at the Geneva Motor Show: the highlight was the 918 Spyder concept car, a high-performance mid-engined sports car with plug-in hybrid technology. With the Cayenne S Hybrid, Porsche presented the first hybrid in series production in the company's history. The third new innovation was the 911 GT3 R Hybrid: the first Porsche racing car with an electric front-axle drive and flywheel generator.

The sporting ambitions of the 918 Spyder were already clear from its prototype.

From 11 to 28 April in Leipzig, Porsche presented the new Cayenne to around 600 journalists from Germany and worldwide. The second generation of the off-road sports vehicle impressed the trade press with outstanding performance and dynamic style.

The second generation of the Porsche Cayenne was also built in Leipzig.

As new basic models in the Gran Turismo series, the Panamera and Panamera 4 were available from May. Both were powered by a newly developed 3.6-litre V6 engine with direct fuel injection and a performance of 220 kW (300 PS). They made their world premiere on 23 April at the Beijing Motor Show.

The Panamera models' engine and drive concepts were enhanced with a 3.6-litre V6 engine.

Early May saw the new top model of the 911 series make its debut appearance: the 911 Turbo S. With 390 kW (530 PS), it produced 30 PS more than the 911 Turbo, accelerated from 0 to 100 km/h in 3.3 seconds and boasted a maximum torque of 700 Newton meters. Consuming just 11.4 litres of fuel over 100 kilometres, it also became the most efficient sports car in its class.

With 530 PS, the 911 Turbo S became the most powerful Porsche 911.

Matthias Müller, new Chairman of the Executive Board of Porsche AG.

In July 2010, Matthias Müller was named Chairman of the Executive Board of Porsche AG with effect from 1 October 2010.

Just a year after production began, the 25,000th Panamera was built in Leipzig.

On 29 July, in Leipzig, the 25,000th Panamera rolled off the assembly line. The ruby-red landmark was delivered to a customer in the US.

The Cayman R was presented at the LA Auto Show in November. With 243 kW (330 PS) and 1,370 kilogrammes of unladen weight, the Cayman R was aimed at buyers with a strong focus on sports capability.

Only 1,621 units of the special edition Cayman R were made.

2011

With the 911 Carrera GTS, Porsche expanded its model range at the Los Angeles motor show. Available as a Coupé and Cabriolet, the GTS was powered by a 300 kW (408 PS) naturally aspirated engine.

With a more powerful engine and extensive standard equipment, the GTS versions were the Carrera model range at its peak.

The Detroit Auto Show saw the 918 RSR concept car make a splash. Known as the "Race Lab", the 918 RSR combined the technology of the 911 GT3 R with the design of the 918 Spyder prototype.

In the 918 RSR, Porsche combined racing technology with the design of a super sports car.

March saw Porsche begin a new chapter in its hybrid strategy with the Panamera S Hybrid. With consumption of 6.8 litres over 100 kilometres (NEDC), the S Hybrid was the most economical Porsche to date.

The Panamera S Hybrid: economical and agile.

The test car Boxster E was the first all-electric Porsche.

As part of a major push towards e-mobility, Porsche tested three all-electric Boxster E cars in spring. The aim was to gather experiences and information in everyday use and regarding usage behaviour.

The new top-range model in the 911 series, the 911 GT3 RS 4.0 proved itself in June: it handled a lap of the Nürburgring north loop in just 7:27 minutes. Available in a limited edition of 600, the car was powered by a 368 kW (500 PS) naturally aspirated engine.

For the first time, a Porsche 911 boasted a four-litre naturally aspirated boxer engine.

After 156 laps and 3,958 kilometres covered, the 911 GT3 RSR from the Manthey Racing team won the 24 Hours of the Nürburgring. Timo Bernhard, Lucas Luhr, Marc Lieb and Romain Dumas were behind the wheel.

The 911 proved its prowess as a rapid and reliable sports car with another victory at the Nürburgring 24 Hours.

On 13 September, Porsche presented the new 911 at the Frankfurt IAA Motor Show. Known internally as the 991, this generation boasted a unique combination of performance and efficiency that set it apart. The seven-gear transmission was a global innovation.

Presented at the 2011 IAA Motor Show: the 911, now in its seventh generation.

2012

February saw the birth of a new generation of the Porsche Boxster. Internally dubbed the 981, it featured a brand-new body, even greater efficiency and idiosyncratic lines.

The Porsche Boxster was fully overhauled for its third generation.

The Panamera model series saw the addition of the especially sporty Panamera GTS. Likewise, the Cayenne became available as a GTS from May. It also offered improved energy performance, more dynamic development of power and a tighter, dropped body.

The GTS versions were a sporty new addition to the Cayenne and Panamera ranges.

On 5 April, the creator of the 911, Ferdinand Alexander Porsche, passed away. After a successful career as an automotive designer, he founded the company Porsche Design in 1972 with his brother Hans-Peter Porsche.

Ferdinand Alexander Porsche died aged 76 in Salzburg.

The Porsche brand was represented from summer onwards at Autostadt Wolfsburg with its own pavilion.

The Porsche Pavilion at the Autostadt in Wolfsburg.

The Panamera Sport Turismo concept car surprised visitors at the Paris Motor Show. As a lifestyle concept car powered by a plug-in hybrid drive, the Panamera Sport Turismo represented a piece of the future.

Porsche surprised the trade press with the Panamera Sport Turismo.

The Los Angeles Auto Show in December was marked by three premieres. It saw the introduction of the new Cayman Generation 981, the 911 Carrera 4 and the Panamera Platinum Edition.

Porsche Cayman and Cayman S, model year 2013.

2013

January 2013 saw the Cayenne S Diesel make its debut. Boasting 281 kW (382 PS), the 4.2-litre eight-cylinder engine with biturbo charging was capable of reaching an 850 Nm torque. NEDC consumption was just 8.3 litres, thanks in part to a stop-start automatic system. The new top model in the Cayenne series was the Turbo S with 405 kW (550 PS).

The Cayenne S Diesel surpassed the Cayenne Diesel in performance and torque.

At the Geneva Motor Show, Porsche showcased the new 911 GT3 from Generation 991 for the first time. This marked the beginning of a new chapter for circuit-ready high-performance sports cars.

The first active rear-axle steering in a Porsche marked a technological milestone.

The active rear-axle steering in the 911 GT3 was a true innovation.

In spring, with the new Panamera S E-Hybrid, Porsche launched the world's first luxury plug-in hybrid. In addition to the Panamera S E-Hybrid, other options such as two Executive versions with a lengthened wheelbase were launched on the market. Another innovation was the brand-new three-litre V6 engine with biturbo charging for the Panamera S and 4S.

The hybrid drive in the Panamera S E-Hybrid was another step towards the electrification of Porsche cars.

Forty years after the first 911 Turbo prototype, Porsche introduced a new generation of the 911 Turbo and Turbo S at the Frankfurt IAA Motor Show. A new all-wheel drive, active rear-axle steering, adaptive aerodynamics, full LED headlights and the six-cylinder biturbo boxer engine delivering up to 412 kW (560 PS) underpinned the new generation's role as a pioneer of technology.

The six-generation Turbo models saw further enhancements in technology and driving dynamics.

On the 50th birthday of the 911, a special-edition 911 model was presented at the IAA Motor Show in Frankfurt. It was set to be available in a limited edition of 1,963 cars.

The first and seventh generations of the Porsche 911.

In June 2013, the 100,000th Panamera was built at the Porsche Factory Leipzig.

The anniversary car is a Panamera S E-Hybrid in Rhodium Silver Metallic.

At the 24 Hours of Le Mans, Porsche scored class wins 99 and 100. Accordingly, the 911 RSR won the GTE Pro class with works drivers Romain Dumas, Marc Lieb and Richard Lietz, while the IMSA Performance Matmut customer team won the GTE Am class.

Another reason to celebrate: the 100th class win at French endurance classic Le Mans.

The highlight of the year was the presentation of the 918 Spyder upon reaching production maturity. Just a few days before the IAA began, the high-performance hybrid sports car set a new record on the Nürburgring north loop of six minutes and 57 seconds.

Porsche handled the Nürburgring north loop in less than seven minutes with the 918 Spyder.

At the same time, Porsche unveiled the new Macan in November at trade fairs held in Los Angeles and Tokyo. The compact SUV is produced at the Porsche factory in Leipzig, where 500 million euros had previously been invested in expanding the site.

The Macan was the third model series to be produced at Porsche Leipzig.

2014

Porsche began 2014 by presenting the new 911 Targa at the Detroit Motor Show. The highlight of this modern classic was its innovative roof system, inspired by the classic Targa frame.

The new 911 Targa brought back the first generation roll bar.

With the 919 Hybrid, Porsche returned to the top category of the sports car world endurance championships. As the fastest research lab and most complex Porsche racing car of all time, the 919 Hybrid was set to take part in all eight rounds of the WEC.

Porsche returned to the WEC prototype class as a factory team with the 919 Hybrid.

The sportiest derivatives of the Boxster and Cayman also bore the GTS initials.

From spring, the mid-engined sports cars Boxster and Cayman were also available in GTS versions. In addition to a boost of 15 PS, the performance of the GTS models was also reflected in details such as a separate front section and a modified rear.

In the 24 Hours of Le Mans, the Porsche team came away from the dramatic closing phase empty-handed. After more than 22 hours, the LMP1 racing car with number 20 and drivers Timo Bernhard, Brendon Hartley and Mark Webber, while in the lead, lost due to a drive defect.

An overhauled generation of the Porsche Cayenne offered more comfort, safety and driving dynamics from September. The Cayenne S E-Hybrid became the world's first plug-in hybrid in the premium SUV segment. It consumed just 3.4 litres over 100 kilometres (NEDC) and was also capable of travelling up to 36 kilometres powered exclusively by electricity.

In November, the series saw the addition of a new Cayenne GTS, whose V6 biturbo delivered 324 kW (440 PS).

With the Cayenne S E-Hybrid, Porsche brought an SUV with a plug-in hybrid drive onto the market.

In the last WEC race of the 2014 season, the Porsche team in São Paulo picked up its first overall win with the 919 Hybrid.

Porsche ended the first racing season for the 919 Hybrid with an overall win in São Paulo.

2015

At the North American International Auto Show in Detroit, Porsche presented two particularly sporty cars in the form of the 911 Targa 4 GTS and the Cayenne Turbo S.

Porsche 911 Targa 4 GTS, model year 2015.

With the Cayman GT4, Porsche presented a new GT sports car in Geneva in early March. The mid-engined sports car's engine, body, brakes and aerodynamics were all designed for maximum driving dynamics. Its technology drew inspiration from the 911 GT3. The 911 GT3 RS was also introduced as the sportiest model with road registration. Its four-litre six-cylinder boxer engine delivered 368 kW (500 PS) and a 460 Nm torque.

The Cayman GT4 handled the north loop in 7:40 minutes.

On 7 May, with One Porsche Drive in Atlanta, in the US state of Georgia, Porsche opened its new customer experience centre with a company headquarters for North America. Unique in the global automotive industry, the location offered office, training and event space in combination with a modular test track, a vintage car gallery and a restoration workshop.

The evolution of the Porsche 919 Hybrid was presented to the public on 26 March at the Circuit Paul Ricard in France. Porsche now entered three 919 Hybrid cars in the WEC races in Spa-Francorchamps and Le Mans. In addition to previous teams Bernhard/Hartley/Webber and Dumas/Jani/Lieb, Earl Bamber, Nico Hülkenberg and Nick Tandy shared the third cockpit.

One Porsche Drive in Atlanta became the main destination for Porsche enthusiasts in the US.

The LMP1 team for the 2015 racing season.

With a double victory, Porsche picked up the 17th overall win at the 24 Hours in Le Mans for the company. Drivers Earl Bamber, Nico Hülkenberg and Nick Tandy were the first to reach the finish line in the 919 Hybrid, exactly 45 years to the day after the first Porsche overall win at the Sarthe.

Porsche won the 24-hour classic on the River Sarthe with the 919 Hybrid.

Porsche took over the toolmaking division of Kuka Systems GmbH with locations in Schwarzenberg in Saxony, Germany, and Dubnica, Slovakia. The division, with over 600 employees, became a wholly owned subsidiary of the Stuttgart sports car manufacturer Dr. Ing. h.c. F. Porsche AG under the name Porsche Werkzeugbau GmbH.

At the IAA Motor Show in Frankfurt, the Mission E concept car enjoyed its world premiere. The four-door car with four single seats offered system performance of over 440 kW (600 PS) and a range of 500 kilometres. Thanks to an 800-volt charging unit developed in-house, the lithium ion batteries are already available once again for 80 per cent of the car's range after just 15 minutes of charging time.

A vision of the future in Frankfurt: the Mission E concept car.

Porsche training centre in Zuffenhausen.

At the start of the new apprenticeship year, Dr. Ing. h.c. F. Porsche AG opened one of the most modern training centres in the entire automotive industry. Covering a total area of 14,000 square metres, the new building on the site of the former tram depot in Zuffenhausen offered space for up to 500 trainees, in addition to students from the Baden-Württemberg Cooperative State University.

In September, Oliver Blume became the new Chairman of the Executive Board of Porsche AG. He replaced Matthias Müller, who moved to Wolfsburg. The Supervisory Board appointed Lutz Meschke, member of the Executive Board for Finance and IT, Deputy Chairman.

Oliver Blume replaced Matthias Müller as Chairman of the Executive Board.

The Porsche Supervisory Board gave the Mission E project the green light. The Zuffenhausen headquarters underwent major refurbishment for production. The company invested around 700 million euros in new facilities, a new paint shop and its own assembly plant. Over 1,000 jobs were created, too.

With the LMP1 team, Porsche secured the world championships in the manufacturer and driver rankings.

Porsche become world champions in the World Endurance Championship. The sports car manufacturer from Zuffenhausen had already won the team ranking before the season's end. As it drew to a close, the driver trio of Timo Bernhard, Brendon Hartley and Mark Webber also took the title in the drivers' ranking after eight rounds.

2016

The 911 R had its world premiere at the Geneva Motor Show. Like its 1967 namesake, the special edition, limited to 991 units and boasting a 368 kW (500 PS) four-litre boxer naturally aspirated engine and a six-gear sports transmission, focused on consistent lightweight construction, maximum performance and an unfiltered driving experience.

A purist in classic garb: the 911 R.

Porsche opened the brand exhibition "Faszination Sportwagen" in Berlin and offered car enthusiasts the opportunity to experience the brand in the capital for the first time.

Porsche turned heads even in the heart of Berlin.

With the 718 Boxster, Porsche drew on the tradition of four-cylinder boxer engines for 718 mid-engined sports cars. In the 1950s and 1960s it racked up numerous racing victories, including at the Targa Florio and in Le Mans. The new engine with turbocharging is more powerful and more efficient at the same time.

The mid-engined Roadster 718 Boxster and 718 Boxster S.

The Supervisory Board of Dr. Ing. h.c. F. Porsche AG decided to found Porsche Digital GmbH. The aim of the Ludwigsburg-based company was to become a leading provider of digital mobility solutions in the premium automotive segment.

Porsche held the world premiere of the new Panamera at Motorwerk in Berlin. The second generation of the Gran Turismo combined the performance of a sports car with the comfort of a luxury saloon. The new biturbo engines are more powerful and – thanks to a new eight-gear dual-clutch transmission – up to 16 per cent more economical.

The second-generation Panamera had a visual and technical overhaul.

Porsche Leipzig GmbH created an additional 600 jobs to produce the new generation of the Panamera, expanding its workforce to around 4,000 highly qualified staff. The company employed people from the local area, making it a major pillar of Saxony as an industrial hub.

The Porsche Leipzig factory following refurbishment.

In Berlin, Porsche opened the Porsche Digital Lab to identify and test innovative information technology. There, teams from Porsche and MHP deal with the question of how innovations from big data and machine learning, micro services and cloud technologies and industry 4.0 and the Internet of Things can become part of practice at Porsche.

2017

Three world premieres saw Porsche make a splash at the Geneva Motor Show. The Panamera Sport Turismo, 911 GT3 and Panamera Turbo S E-Hybrid set benchmarks in powerful sustainability, sporting performance and speedy variability.

The Panamera Sport Turismo featured a 4+1 seat concept.

New wholly owned subsidiary: Dr. Ing. h.c. F. Porsche AG took over 35 per cent of shares in Porsche Lizenz- und Handelsgesellschaft mbH & Co. KG (Porsche Design Group) from a Porsche family investment company. The Stuttgart sports car manufacturer already held 65 per cent of shares.

On 11 May, the one-millionth Porsche 911 left the factory. In the individual colour "Irish Green" and with numerous exclusive special features such as Pepita seats and a handmade mahogany wooden steering wheel, this one-off recalled the first 911 from 1963. The anniversary Coupé was based on the 911 Carrera S with the more powerful 331 kW (450 PS) boxer engine and seven-gear transmission. It accelerated from 0 to 100 km/h in 4.1 seconds and achieved peak speeds of 312 km/h.

The one-millionth 911 in New York City.

Porsche picked up an overall victory for the third time running at the 24 Hours of Le Mans.

Le Mans 2017: hat-trick for the 919 Hybrid. On Saturday evening at 6.30 p.m., for the Porsche with starting number 2 from Earl Bamber, Timo Bernhard and Brendon Hartley, the 24 Hours seemed lost: a front-wheel drive missing and over an hour of repair time. But the 85th edition of the endurance classic in Le Mans took such dramatic twists and turns that an almost superhuman come-back from 56th place and an 18-lap deficit was rewarded with an overall win. Thou-sands of spectators watched one of the most thrilling 24-hour races of all time at the Porsche Museum in Zuffenhausen.

At the Frankfurt International Motor Show, the new Cayenne took centre stage at the Porsche booth. However, the world premiere of the new generation took place on 29 August with an especially composed symphony on the roof terrace of the Porsche Museum.

The third-generation Cayenne.

With the foundation of the joint venture IONITY, the BMW Group, Daimler AG, Ford Motor Company and the Volkswagen Group with Audi and Porsche led the way for building the most powerful high-speed charging networks for electric cars in Europe.

Details of the fuel consumption and CO_2 emissions can be found on page 282 and 283.

In 2017, Porsche took the triple at the FIA World Endurance Championship for the third year running: victory in Le Mans, a win in the manufacturers' world championship and a successful defence of the drivers' title. Thus ended the involvement of Porsche in the WEC's top category after four years.

Celebrations after the 24 Hours of Le Mans.

2018

At the Geneva Motor Show, Porsche presented its new high-performance sports car, the 911 GT3 RS. With its racing-car chassis and four-litre high-rev naturally aspirated engine with 383 kW (520 PS), it accelerated from zero to 100 in just 3.2 seconds and reached maximum speeds of 312 km/h.

The Porsche 911 GT3 RS completed a lap on the Nürburgring in 6:56.4 minutes.

The second highlight of the Geneva Motor Show was the Mission E Cross Turismo concept car, an all-electric cross utility vehicle (CUV). The four-door Cross Turismo featured a thrilling design with striking off-road elements and an innovative display and control concept with a touchscreen and eye-tracking control.

Concept car Mission E Cross Turismo.

Porsche took another step on the path to e-mobility by presenting the new Cayenne E-Hybrid. It combines driving dynamics with maximum efficiency and delivers a system performance of 340 kW (462 PS).

Porsche further expanded its hybrid range with the Cayenne E-Hybrid.

Three-time WEC winner 919 Hybrid said farewell with a tribute tour as 919 Hybrid Evo. Without the strict rules-based limitations of the racing series, the hybrid drivetrain achieved system performance of 1,160 PS. It helped Neel Jani set a new record time on the Grand Prix course in Spa. In June, Timo Bernhard achieved the new record time of 5:19.6 minutes on the Nürburgring.

New best time at the Nürburgring.

Details of the fuel consumption and CO_2 emissions can be found on page 282 and 283.

Porsche steps up its social responsibility: the Ferry Porsche Foundation was founded to mark 70 years of the brand's existence. Its aim: to support projects in education and social affairs and to help children and young people. The foundation started life with ten million euros' worth of funding, with another ten million euros set to follow in a second phase.

The 911 Speedster Concept was presented as part of the festivities marking 70 years of Porsche as the ready-to-drive prototype for a particularly exciting sports car. Its design recalled historic forebears such as the 356 Speedster. With its purist design, the 911 Speedster Concept reflected the core of the Stuttgart sports car manufacturer's brand in precise clarity.

The Ferry Porsche Foundation receiving the certificate of recognition.

The drive technology of the 911 Speedster Concept was taken from the GT models from the Porsche Motorsportzentrum.

In the GT Pro class at 24 Hours of Le Mans, the Porsche 911 RSR with start number 92 came in first at the 86th edition of the classic endurance race after 344 laps. The works car drew on the famous "Pink Pig" design of the 917/20 from 1971. Porsche also won the second GT class.

The 911 RSR scored a class win at Le Mans 2018.

Details of the fuel consumption and CO_2 emissions can be found on page 282 and 283.

At the historic motorsporting event Rennsport Reunion, Porsche presented the new 935 at the Laguna Seca Raceway. The car was not homologated, allowing it to operate outside the usual regulations. The body of the 515 kW (700 PS) racing car recalled the Porsche 935/78. Like its predecessor,

The Porsche 935 was sold out before start of production.

large parts of the car's body consisted of carbon-fibre-reinforced polymers. With its streamlined rear, the 935 spanned a length of 4.87 metres.

On 27 November the new 911, known internally as the Generation 992, celebrated its world premiere in Los Angeles. According to Chairman of the Executive Board Oliver Blume: "the eighth generation of the 911 is even more powerful, more exciting and more efficient than its predecessor. Despite all the innovations, the 911 remains what it was in the very beginning: a pure sports car, and the beating heart of Porsche. Our icon."

World premiere of the eighth 911 generation in Los Angeles.

2019

Porsche opened the 89th Geneva Motor Show with three sporting innovations: the highlight among them was the new 911 Cabriolet. The mid-engined 718 T was presented as a Boxster and Cayman, delivering 300 PS. The Macan S with its new V6 turbo front engine completed the trio of debuts.

The Cayenne family grew: on 21 March, the world premiere of the Cayenne Coupé took place in Stuttgart. The Coupé version had an even more dynamic appearance thanks to a roof line that sloped backwards much earlier, positioning it as the sportiest model in the segment. Porsche Engineering took charge of overall vehicle development. Simulation and virtual development tools were primarily used to develop the car and ten digital prototypes were created. It was launched on the market in late May.

Thanks to new hydraulics, the opening time of the top with its set glass rear window was reduced to around twelve seconds.

The large glass roof created a unique feeling of space in the Cayenne Coupé.

Details of the fuel consumption and CO_2 emissions can be found on page 282 and 283.

Top performance online: Porsche expanded its brand family of motorsporting competitions to include a virtual championship, the Porsche Esports Supercup. The name Supercup referred to the top category of manufacturers' cups and highlighted the international spirit and relevance of the online racing series, set up by the sports car manufacturer in collaboration with iRacing.

Porsche became the new world champion at the FIA World Endurance Championship (WEC): Porsche secured the title in the GTE Pro class with a race to spare at the end of the "Super Season 2018/2019". With the 109th class win at the 24 Hours of Le Mans, the Project 1 team won the GTE Am class and the driver and team world championships at the same time.

Class win in Sarthe.

The Goodwood Festival of Speed saw the new 911 RSR make its global debut. In line with FIA GTE regulations, the racing car was a brand-new development. Its centrepiece was the high-efficiency six-cylinder naturally aspirated engine installed in front of the rear axle, with a capacity of 4,194 cubic metres. Depending on classification, it was capable of delivering 515 PS. It was the biggest boxer engine ever to be fitted in a Porsche 911 at the factory. It offered even better drivability over a wider engine speed range than the proven four-litre unit of its predecessor. The 911 RSR made its racing debut in the UK as well – at the FIA World Endurance Championship (WEC) season start on 1 September in Silverstone.

Porsche 911 RSR with 4.2 litre displacement.

Porsche entered Formula E in November with the 99X Electric. The development of the power train was crucial to the competition. For the purposes of cost control, Formula E provided the car's chassis and the unit battery, while all drive components were designed by the manufacturer. This allowed Porsche to take its own approach to developing key technologies. This included the electric motor, converter, a brake-by-wire system, transmission, differential, driveshafts, the supporting structure and chassis parts on the rear axle, as well as the cooling system and control unit. Neel Jani and André Lotterer were regular drivers for the first season. The Porsche 99X Electric had already debuted worldwide to almost a million viewers in a live stream on Twitch on 28 August.

During an extensive test programme in Spain, the 99X Electric covered more than 1,000 kilometres in three days.

Details of the fuel consumption and CO_2 emissions can be found on page 282 and 283.

With a spectacular world premiere occurring simultaneously on three continents in Toronto, Berlin and Fuzhou, Porsche presented its first all-electric sports car, the Taycan, on 4 September. The name, with a meaning analogous to "vibrant young horse", took linguistic inspiration from eastern Asia and drew on the leaping steed in the Porsche emblem. The system performance was over 440 kW (600 PS), enabling acceleration from zero to 100 in under 3.5 seconds. Unlike competing cars, the Taycan was capable of accelerating up to ten times in a row – without losing power. The Taycan was also "best in class" when it came to recuperation management. The central placement of the battery on the car's underbody gave it a particularly low centre of gravity, making it ideal for navigating bends and turns – just like the 911.

The Porsche Taycan at the world premiere at the Niagara Falls.

On 9 September, Porsche opened the new Taycan factory at its Zuffenhausen headquarters with the slogan "smart, lean and green". It was joined by 320 guests from the world of politics, business and the media. This marked the beginning of a new era of car production for the company. Manufacture was connected and flexible. Instead of a classic assembly line, Porsche used driverless transport systems that moved components and cars from station to station.

Driverless transport systems are used for the assembly of the electric drives and axles of the Taycan.

Porsche heralded the 75th brand anniversary with a special exhibition in Berlin.

2020 — 75 YEARS OF PORSCHE SPORTS CARS — 2023

"Porsche has only remained Porsche by constantly changing."

Oliver Blume

1930 1940 1950 1960 1970 1980 1990 2000 2010 2020s

2020

Porsche expanded its mid-engined sports car range: the 718 Cayman GTS 4.0 and 718 Boxster GTS 4.0 went on sale in March 2020. The combination of GTS equipment and a four-litre six-cylinder boxer engine allowed them to combine tight sporting performance and everyday comfort. Two GTS models were added to the SUV range, too: as the Cayenne GTS and Cayenne GTS Coupé, they improved their driving performance in all disciplines with the four-litre biturbo V8 engine delivering 338 kW (460 PS).

Sporty double with a four-litre capacity.

After the 90th Geneva Motor Show was cancelled, Porsche presented the new 911 Turbo S from Generation 992 at a virtual press conference. The top-range 911 delivered 478 kW (650 PS) and appeared more powerful, dynamic and comfortable than ever. Its centrepiece was a new 3.8-litre boxer engine with two VTGs. The maximum torque now reached 800 Newton metres and the sprint from zero to 100 km/h was cut to 2.7 seconds with the Turbo-specific eight-gear PDK transmission. The 911 Turbo S was available as a Coupé and Cabriolet from the start.

Porsche 911 Turbo S.

Details of the fuel consumption and CO_2 emissions can be found on page 282 and 283.

After almost two months at a standstill due to the Covid-19 pandemic, production gradually revved up again at the Zuffenhausen headquarters and Leipzig plant in May. The Porsche Museum also reopened subject to pandemic restrictions from the federal government and state of Baden-Württemberg.

Standstill in the Porsche factory in Zuffenhausen.

For the first time, Porsche presented a new model in its own Web TV format in mid May: the 911 Targa. One remarkable feature was the fully automatic roof system with its distinctive wide arch and moving roof section above the front seats. The models were powered by a six-cylinder boxer engine with a capacity three litres and biturbo charging.

World premiere on web TV: the Porsche 911 Targa.

As the first of four collector's items from the Heritage Design strategy, Porsche presented the 911 Targa 4S Heritage Design Edition in early June. Historic design elements in the interior and exterior were reinterpreted and combined with the latest technology in an exclusive special model from the Porsche Exclusive Manufaktur – with a limited edition of 992 cars.

The 911 Targa 4S Heritage Design Edition cited elements of the 1950s and 1960s.

In a spectacular season finale, the two Porsche 911 RSR cars achieved a double victory at the 12 Hours race in Sebring. The thrilling and extraordinarily successful farewell to the hotly contested GTLM class with many manufacturers in North America marked the end of a programme that delivered an incredible record after seven years of use: Porsche achieved seven title wins in the IMSA series, five championships in the North American Endurance Cup, 21 race victories and 56 podium places.

The 911 RSR said farewell to the GTLM class in the USA after seven successful years.

Details of the fuel consumption and CO_2 emissions can be found on page 282 and 283.

The 911 Generation 992 Porsche 911 GT3 Cup started its first motorsport season in 2021.

The new manufacturers' championship racing car Porsche 911 GT3 Cup did its last few laps in secret: the next competition car for the Porsche Mobil 1 Supercup and selected national Carrera Cups was the first racing car based on the 911 Generation 992.

2021

Porsche marked the Roadster family's 25th birthday with a special anniversary model: the Boxster 25 Years. The special edition, limited to 1,250 cars worldwide, was based on the 294 kW (400 PS) Boxster GTS 4.0 with a four-litre six-cylinder boxer engine. It drew on many design features from the Boxster prototype that heralded the success story of the open-top two-seater in 1993 at the Detroit Motor Show. The Boxster finally went into series production in 1996.

For the anniversary model, Porsche reinterpreted the Neodyme colour of the 1993 Boxster design study.

The new 911 GT3 was an even more efficient application of pure racing technology to a series model. The double wishbone front axle layout and the sophisticated aerodynamics with swan-neck rear wing came from the 911 RSR. The 911 GT3 R served as a basis for the 375 kW (510 PS) four-litre six-cylinder boxer engine.

Racing technology for series production.

The world premiere of the new Taycan Cross Turismo took place on 4 March 2021. As the brand's first cross utility vehicle, the five-door Porsche combined classic driving dynamics with a high-tech chassis suitable for off-road use. The CUV featured all-wheel drive, air suspension and Performance Battery Plus as standard.

The Taycan Cross Turismo is the all-rounder of Porsche's electric sports cars.

The Porsche Mobil 1 Supercup used renewable fuels at the start of the season: as a first step, the international manufacturers' tournament used advanced organic fuels and "eFuels". To gain experience for potential further developments with regard to future use in serially produced cars, Porsche and its project partner ExxonMobil tested the suitability of renewable liquid fuels under tough race-driving conditions. The new Porsche 911 GT3 Cup with its high-speed, high-performance six-cylinder boxer engine was the ideal test lab for the Esso Renewable Racing Fuels: its powerful 4.0-litre unit had already been tested for use with renewable energies.

Porsche capped off the Cayenne model series mit with a top-of-the-range sports car: consistently tuned to maximum longitudinal and lateral performance, the new Cayenne Turbo GT combined extraordinary driving dynamics with extensive everyday road capability. The foundation for this was the 471 kW (640 PS) four-litre V8 biturbo engine. With an even sportier design, the Cayenne Turbo GT, available exclusively as a four-seater Coupé, was equipped as standard with all available chassis systems and specially designed performance tyres.

The Cayenne Turbo GT was in dealers-hips from mid September.

Details of the fuel consumption and CO_2 emissions can be found on page 282 and 283.

Porsche AG assumed the role of strategic shareholder in the Bugatti–Rimac joint venture.

Porsche and Rimac agreed on the establishment of a joint venture involving Bugatti. Oliver Blume and Mate Rimac signed the relevant paperwork and announced the name of the new hypercar manufacturer: Bugatti-Rimac. The joint venture based in Zagreb, Croatia, was founded in the fourth quarter of 2021.

Rimac held 55 per cent and Porsche 45 per cent of shares in the Bugatti-Rimac joint venture. Porsche also held a direct stake of 24 per cent in Rimac. Bugatti was brought into the joint venture through the current owner, Volkswagen. The shares were then transferred to Porsche.

Classic colours made a comeback for all Porsche models. With the new options "Paint to Sample" and "Paint to Sample Plus", the Porsche Exclusive Manufaktur expanded the range of special exterior colours for all model series. The focus was the availability of classic Porsche colours already familiar from the company's history. There was also the option to create shades based on customer requirements for highly specific colour preferences.

The Porsche Exclusive Manufaktur cited a photo of twelve 911 Carrera RSR 3.0 before loading for the 1973 International Race of Champions.

Motorsports partner Manthey Racing achieved its seventh overall victory at the Nürburgring 24 Hours. Works drivers Kévin Estre from France, Michael Christensen from Denmark and Matteo Cairoli from Italy finished first at the 49th edition of the classic endurance race behind the wheel of a Porsche 911 GT3 R. This success capped off celebrations marking 25 years since the foundation of the Eifel-based racing stable. Manthey continued its record as the most successful team with a seventh win at the 24 Hours. It was also the 13th triumph for a Porsche.

Overall win at the Nürburgring to mark the team's anniversary.

Details of the fuel consumption and CO_2 emissions can be found on page 282 and 283.

Porsche offered a glimpse of motoring's future with the Mission R concept car at IAA Mobility 2021 in Munich. The progressive all-electric racing car combined the latest technology and sustainable materials such as natural fibre reinforced plastics, paired with a passion for motorsport. The two electric engines delivered up to 800 kW (1,088 PS) on the racetrack in "qualifying mode". The battery capacity of around 80 kWh and innovative recuperation enabled sprint racing without loss of performance.

The Porsche Mission R was the vision for an all-electric car for customer motorsport.

At the Los Angeles Auto Show, Porsche presented the new 718 Cayman GT4 RS Clubsport. The mid-engined circuit sports car from Weissach was based on the equally new series model 718 Cayman GT4 RS. Its most striking feature was the 4.0-litre six-cylinder boxer engine. This resembled the high-rev unit from the 911 GT3 Cup and delivered 368 kW (500 PS).

High-performance customer racing car for the GT4 class.

2022

For the first time, Porsche launched a four-door sports car with the suffix T.

In February, Porsche presented the Macan T. The initial "T" stood for touring and this marked the first time it was bestowed on a four-door sports car. The Macan enabled the comparatively pure driving experience typical of "touring" cars, thanks to the reduced-weight two-litre turbo engine, ensuring high starting performance, and the sporty standard equipment.

The new Porsche 911 Sport Classic, a homage to Porsche design from the 1960s and 1970s, went on the market in April. As part of the Heritage Design Edition, Porsche presented the second in a series of four small-scale productions in the shape of the 911 Sport Classic. The limited edition of 1,250 units was developed by the Porsche Exclusive Manufaktur in collaboration with the "Style Porsche" design department. The models' appearance was inspired by the first generation of the 911 from 1963 and the 911 Carrera RS 2.7 from 1972.

Ducktail times two: 911 Carrera RS 2.7 from 1972 and 911 Sport Classic from 2022.

Details of the fuel consumption and CO_2 emissions can be found on page 282 and 283.

At the 24 Hours of Le Mans, Porsche won the GTE Pro class with the 911 RSR. On 12 June, works drivers Gianmaria Bruni from Italy, Richard Lietz from Austria and Frédéric Makowiecki from France took first place after over 350 laps. In this last works appearance for the 911 RSR, the team set a new distance record in the GTE class of 4,769 kilometres.

The 911 RSR scored the 110th Porsche class win at the 24 Hours of Le Mans.

Designed for maximum performance, the Porsche 911 GT3 RS was launched in August. The four-litre high-rev naturally aspirated engine delivered 386 kW (525 PS), reached a top speed of 296 km/h and accelerated from zero to 100 km/h in 3.2 seconds. The street-legal high-performance sports car took advantage of technologies and principles from

Porsche 911 GT3 RS.

motorsport time and again: thanks to a large central cooling unit in the boot area, there was more space for aerodynamic elements that increase overall performance.

Porsche began a new era as a listed company. Since 29 September 2022, the preferred shares have been traded on the regulated market of the Frankfurt Stock Exchange (Prime Standard) under the ticker symbol "P911", the WKN number "PAG911" and the ISIN "DE000PAG9113". As of 19 December 2022, Porsche AG had joined the DAX as a "fast entry".

Porsche completed its initial public offering.

At the Los Angeles Auto Show in November, Porsche unveiled a very special version of the iconic 911 sports car: the new 911 Dakar. This was the first two-door series sports car to offer outstanding off-road features. The model name refers to the first Porsche overall win at the Paris–Dakar endurance rally in 1984, where Porsche used an all-wheel drive in motorsport for the first time. The triumph of the 911 Carrera 3.2 4x4 Paris-Dakar (953) was also the first win for a sports car in the desert marathon.

Porsche 911 Dakar.

With a double victory at the Mexico City E-Prix the TAG Heuer Porsche Formula E team celebrated its biggest success to date. In the Autodromo Hermanos Rodriguez, Pascal Wehrlein took the first win for Porsche in pole position at the ABB FIA Formula E world championship in a Porsche 99X Electric. His teammate André Lotterer finished second, capping off the

Details of the fuel consumption and CO_2 emissions can be found on page 282 and 283.

Stuttgart sports car manufacturer's success in the third Formula E race of the season.

Double victory for Porsche at Formula E.

The first DTM triumph for Porsche with Team 75 Bernhard: led by Porsche brand ambassador Timo Bernhard, Team 75 Bernhard and works driver Thomas Preining picked up the first win for Porsche at the DTM, a racing series with a long tradition. His Norwegian teammate Dennis Olsen followed in second place. Works driver Laurens Vanthoor came fourth.

Thomas Preining and Team 75 Bernhard scored the first DTM-win for Porsche.

At Goodwood in the UK the Porsche Penske Motorsport team presented the new Porsche 963 in classic Porsche motorsports colours: white, red and black. The hybrid prototypes according to LMDh regulations are set to start in 2023 in the FIA World Endurance Championship WEC and the IMSA WeatherTech SportsCar Championship in the world's greatest endurance races.

Hybrid race car Porsche 963.

With the newest generation of the 911 GT3 R, Porsche presented a new customer racing car. The challenger for worldwide use in GT3 series was based on the 911 Generation 992. The developers focused primarily on improved drivability for professionals and "gentleman drivers" and bigger power reserves depending on classification through the "Balance of Performance" (BoP). Optimised running costs and a racing car that was easier to use for the team were also prioritised.

The Porsche 911 GT3 R featured an especially lightweight body made of carbon.

The Porsche Factory Leipzig celebrated its anniversary: in 2002, or twenty years ago to be precise, the first car – a Porsche Cayenne – rolled off the assembly line as part of factory opening celebrations. What started out small became the beginning of a Porsche success story. The first assembly plant underwent several expansions to become a full-service plant and is also home to a Porsche Experience Centre. Today, the production site in Saxony is one of the region's biggest employers with over 4,400 employees – and an economic powerhouse for the city.

20 Years of Porsche Leipzig.

Experiencing Porsche in Stuttgart city centre.

Porsche Germany opened the world's first brand store in the heart of Stuttgart city centre in October. The result was a place where people could experience the brand and its core philosophy, "Driven by Dreams", first-hand.

The world's first integrated commercial facility der for the production of eFuels went into operation. The facility was set up by sports car manufacturer Porsche in collaboration with its Chilean partner HIF, Siemens Energy and a range of other international companies in Punta Arenas in Chile with the aim of producing fuel that is potentially almost carbon-neutral (eFuel). Attendees at the opening of this future-oriented project included Chilean Energy Minister Diego Pardow, Chilean Economy Minister Nicolás Grau and Baden-Württemberg's Minister of Transport Winfried Hermann. The fuel produced

Porsche focuses on renewable fuels.

was initially intended for use in flagship projects and Porsche Experience Centres. After the pilot stage, a major scaling of production was planned to begin in the middle of the decade.

2023

The Porsche Vision 357 exterior prototype was the highlight of the special exhibition "Driven By Dreams. 75 Years of Porsche Sports Cars", which opened in late January at "DRIVE. Volkswagen Group Forum" in Berlin. The prototype on the technical platform of the 718 Cayman GT4 RS was a homage to the Porsche 356. With its monolithic shape, the narrow passenger cell with the flyline sloping early to the rear, and its broad shoulders, the proportions of the Porsche Vision 357 recall its ancestor. The windscreen wrapped sharply around the A-pillars. At the same time, it showed possible directions of a future design philosophy. A striking example was the light signature, which has been consistently developed further.

Premiere of the Porsche Vision 357 at the opening ceremony in Berlin.

The new hybrid prototype Porsche 963 gained valuable experience in its racing debut at the IMSA WeatherTech SportsCar Championship. Both cars dropped out of running in the 24 Hours of Daytona due to technical defects. However, after undergoing repairs, some of which were extensive, the two driver crews picked up some important points in the North American championships.

The Porsche 963 took to the starting line in Daytona for the first time.

Details of the fuel consumption and CO_2 emissions can be found on page 282 and 283.

A strong start for Porsche in the Formula E season: the new Porsche 99X Electric Gen3 picked up a double victory at the Mexico City E-Prix. Porsche customer team Avalanche Andretti and its driver Jake Dennis (GBR) won the opening race at the ABB FIA Formula E world championship in Mexico City. For the TAG Heuer Porsche Formula E team, works driver Pascal Wehrlein (GER), who won in Mexico in 2022, came in second after a remarkable comeback.

Double win to kick off the season.

Dr. Wolfgang Porsche turned 80 on 10 May. As the Chairman of the Supervisory Board and Porsche family spokesman, he is following in the footsteps of his father and grandfather. The face of the brand, he is its biggest ambassador.

Dr Wolfgang Porsche.

On 8 June, Porsche marked its 75th brand anniversary with a celebration in Stuttgart-Zuffenhausen and the Festival of Dreams at Hockenheimring the following weekend. The motto of the celebrations was "Driven by Dreams". This reflects the awareness and approach the team behind Porsche take every day as they work to make dreams come true for customers around the globe. Oliver Blume: "For 75 years, Porsche has stood for passion, dreams and a pioneering spirit. We're celebrating with people who are driven by dreams."

Porsche celebrated its 75th brand anniversary.

PORSCHE
75
357

Evolution of the Porsche 911

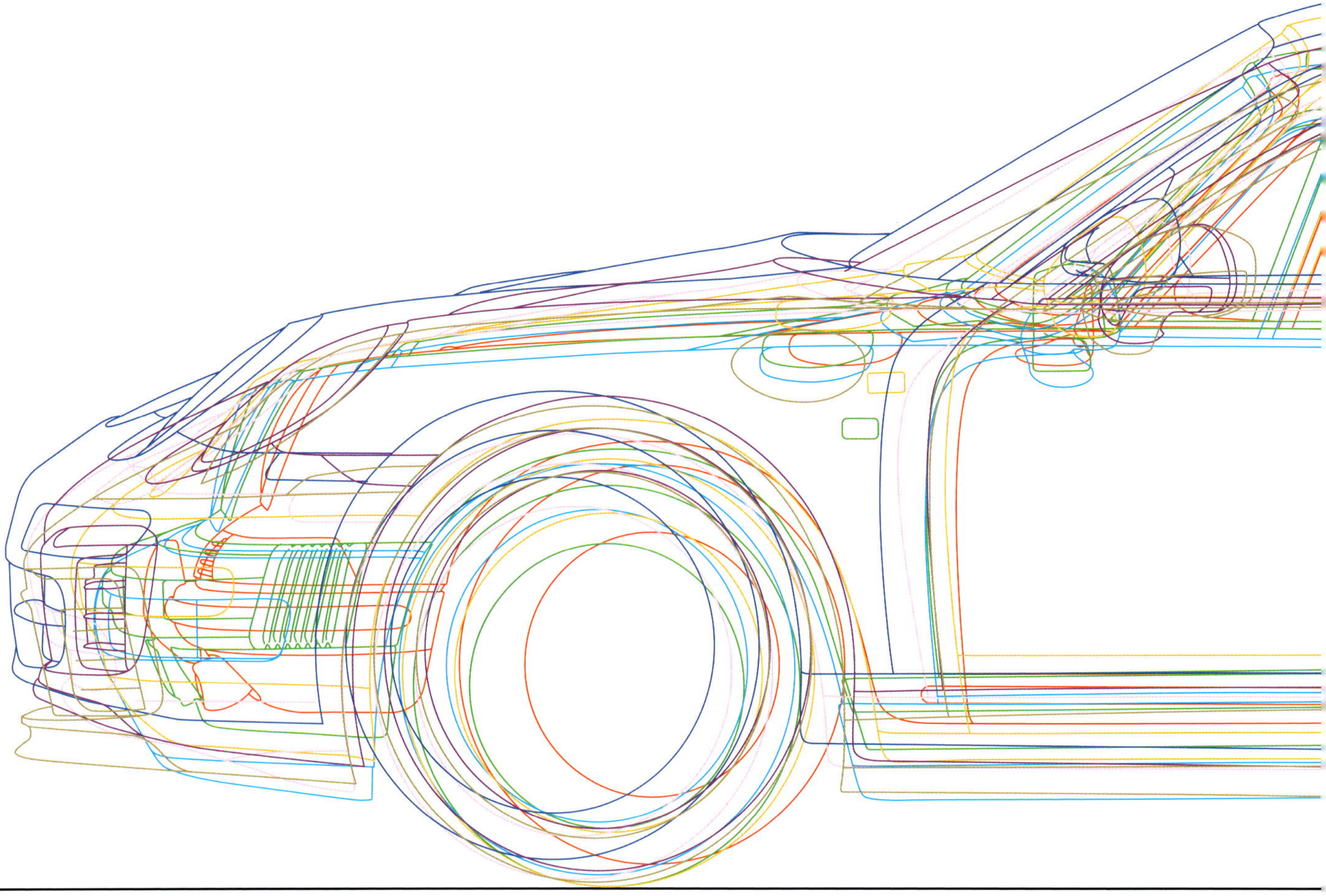

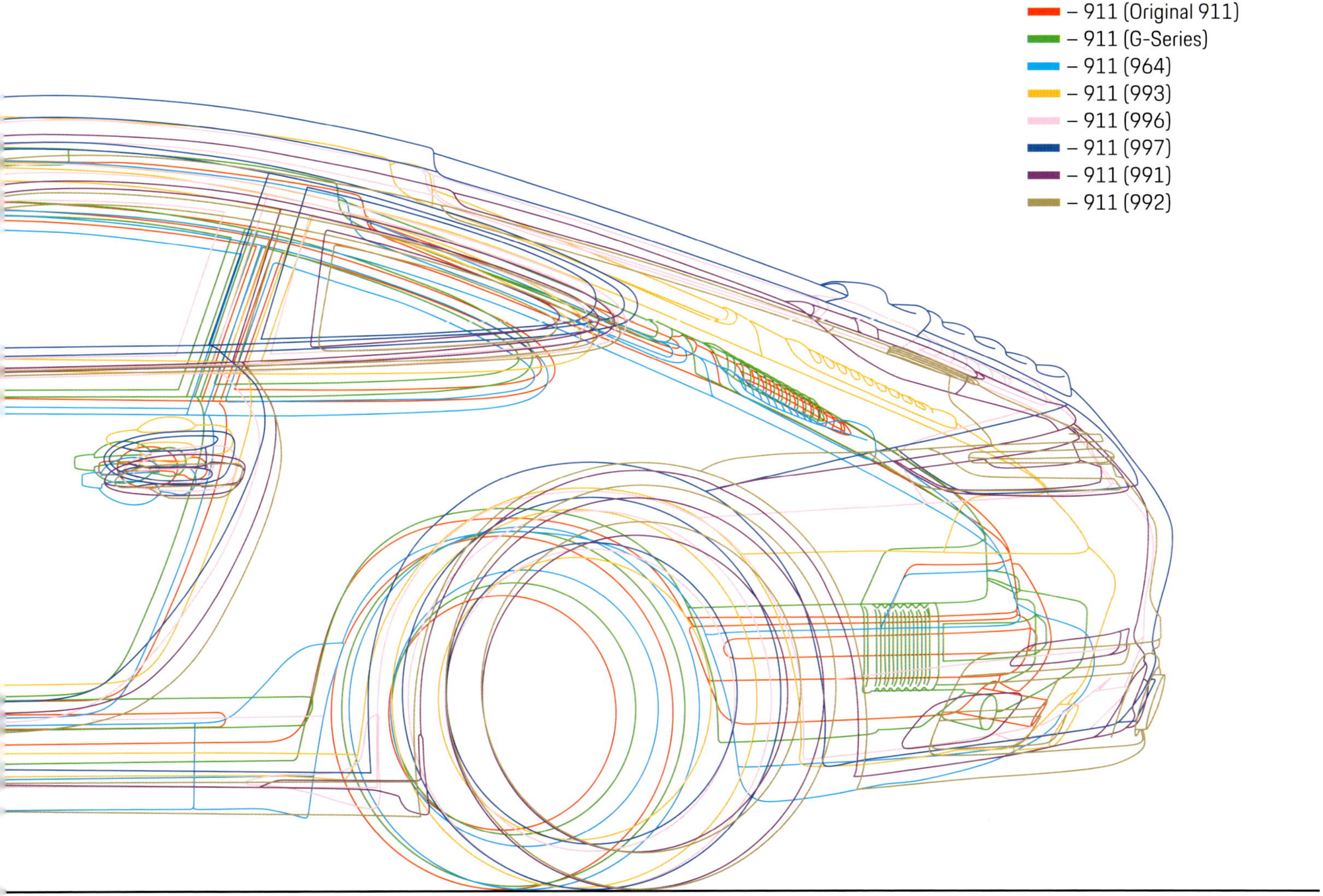
– 911 (Original 911)
– 911 (G-Series)
– 911 (964)
– 911 (993)
– 911 (996)
– 911 (997)
– 911 (991)
– 911 (992)

Generations of the Porsche 911 Turbo

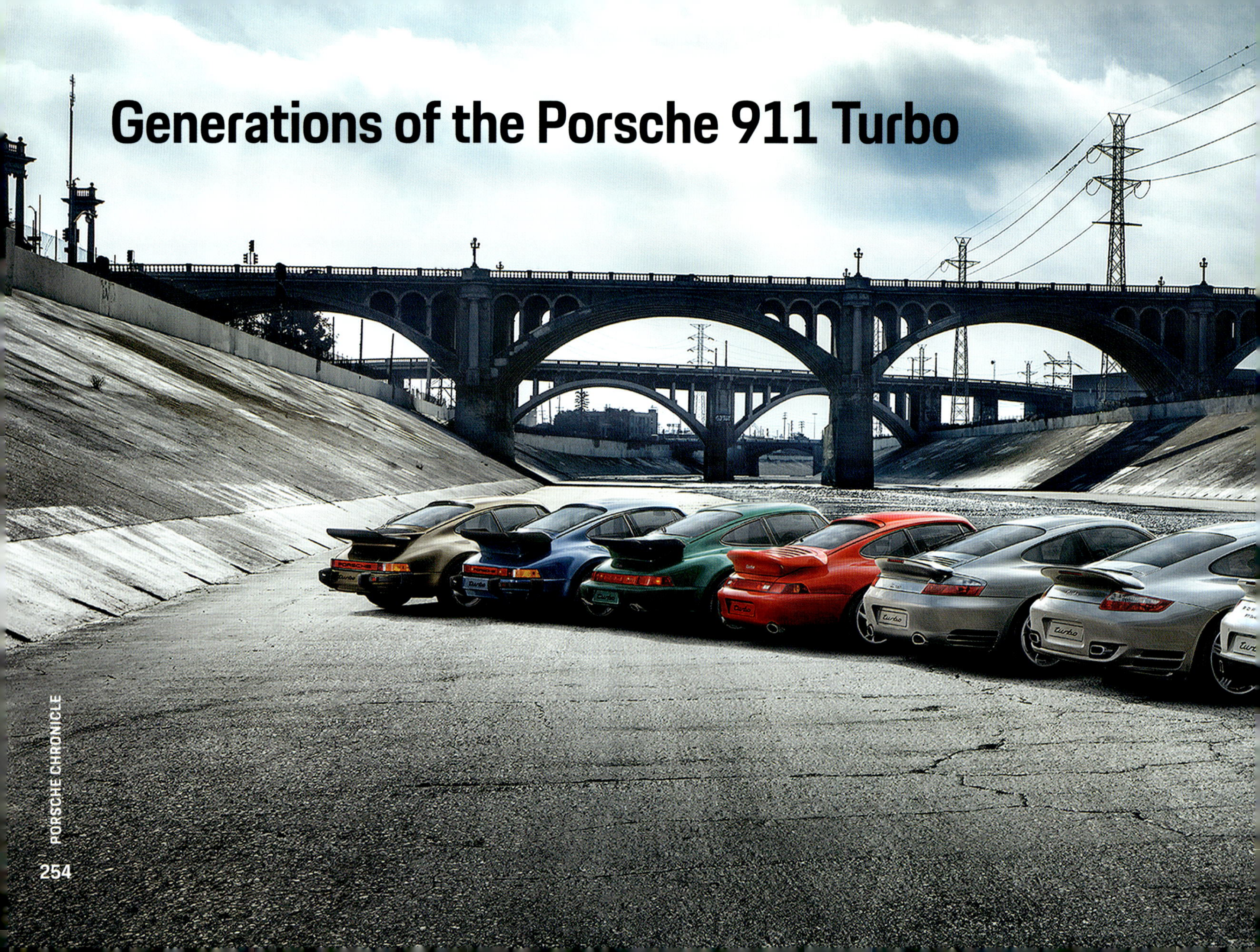

Details of the fuel consumption and CO_2 emissions can be found on page 282 and 283.

The Porsche success story at Le Mans

WINNER IN ALL CLASSES

Le Mans is unique. There is the track length of 13.6 kilometres with four straights – each as long as the longest straight in Formula 1. Then there is the race distance. And the starting field: around 55 cars compete in the world-famous race every year. The fastest cars – usually specially developed prototypes – fight for the overall victory. Depending on displacement and technical design, sports cars and Gran Turismo race cars also compete in additional classes. Porsche holds a record here as well with 108 successes (as of June 2021). The story of the race cars from Stuttgart-Zuffenhausen in Le Mans started with a series of class victories of the 1.1 to 1.6-litre Porsche cars in the years 1951 to 1958.

The factory and customer teams repeatedly drove to decisive victories. In 1970, the 917, 908 and 911 won all classes, and the 914 four of five; this feat of victory in all classes was repeated in 1982 with the 956, 935, 934 and 924. After the first class victory of a Porsche 911 in 1966, the sports car icon regularly finished in the leading positions, above all among the near-production race cars.

Porsche 917 KH

Richard Attwood

Hans Herrmann

1970

Quantum leap

A powerful V12 engine and an extremely light race car: Porsche achieved the first overall victory in Le Mans with a quantum leap.

Porsche 917 KH

Dr. Helmut Marko

Gijs van Lennep

1971

Who dares wins

The light magnesium frame of the white 917 short-tail was a daring choice. The engine was more powerful than in the previous year.
The record distance of the winners remained unbeaten for decades.

Porsche 936/76

Jacky Ickx

Gijs van Lennep

1976

Turbo triumph

Porsche used a turbo engine in Le Mans for the first time in 1974. Two years later, the 936 Spyder was the first turbo winner in the 24 Hours of Le Mans.

Porsche 936/77

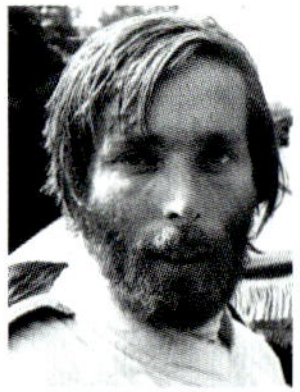

Jürgen Barth

Hurley Haywood

Jacky Ickx

1977

Nine laps behind and victory

The race was just one hour old and the Porsche 936/77 Spyder was in 41st place after a repair. 23 hours later, the Spyder crossed the finish line on five cylinders – as the winner.

Porsche 935 K3

Klaus Ludwig

Don Whittington

Bill Whittington

1979

Milestone

The German Porsche Kremer Racing Team established a milestone with the first win for a Porsche customer and the first success of a rear-engine race car.

Porsche 936/81

Jacky Ickx

Derek Bell

1981

Start of a winning run

The 936 Spyder was already in the museum when Porsche decided to compete in Le Mans. With a new engine and new aerodynamics, the 936/81 started the record series of seven triumphs in succession.

Porsche 956 LH

Jacky
Ickx

Derek
Bell

1982

Debut win

An experienced and well-functioning team, a proven technology package and revolutionary aerodynamics proved to be an unbeatable combination.

Porsche 956 LH

Al
Holbert

Hurley
Haywood

Vernon
Schuppan

1983

Nine of ten

The Porsche 956 quickly became the most successful prototype. Porsche now also built the race car for customers. The result: victory number eight. And nine 956 cars in the top ten.

Porsche 956 LH

Henri
Pescarolo

Klaus
Ludwig

1984

First win for customer 956

The Porsche 956 remained the car to beat. Joest Racing won the race, and seven more 956 cars finished in the top ten. The works team did not compete in protest against a change in the rules.

Porsche 956 LH

Paolo Barilla

Klaus Ludwig

John Winter

1985

Works team beaten, Porsche success celebrated

That is also successful customer motorsport: the private Joest Racing Team and the Richard Lloyd Racing Team finished the race ahead of the 962 C of the works team.

Porsche 962 C LH

Derek Bell

Hans-J. Stuck

Al Holbert

1986

That was still missing

The Porsche 962 was almost unbeatable, driving to race victories and championships. And now it also achieved victory in Le Mans.

Porsche 962 C LH

Derek Bell

Hans-J. Stuck

Al Holbert

1987

Reward for a high-risk strategy

After just one hour, there was just one works Porsche 962 C left in the race against very strong British competition. The team decided on a strat-egy of all-out attack – and triumphed.

Porsche 962 Dauer Le Mans GT

Mauro
Baldi

Yannick
Dalmas

Hurley
Haywood

1994

Renaissance of the 962

The era of the famous Porsche 962 C drew to a close at the end of the 1980s. The former serial winner returned in a Gran Turismo version and won in its only race.

TWR-Porsche WSC Spyder

Manuel
Reuter

Alexander
Wurz

Davy
Jones

1996

Goodbye America, bonjour Le Mans

Porsche designed the TWR-Porsche WSC Spyder for the US, but stopped the programme. Joest Racing borrowed the race car, and Weissach perfected the aerodynamics. The Spyder won.

TWR-Porsche WSC Spyder

Michele
Alboreto

Stefan
Johansson

Tom
Kristensen

1997

Repeat offender

Joest Racing took over the victorious TWR-Porsche WSC Spyder from 1996, invested all its hopes in this one race car, and was again triumphant.

Porsche 911 GT1 '98

Allan McNish

Stéphane Ortelli

Laurent Aïello

1998

Final sprint and double victory
A race of the century with five works teams on the starting grid. In the final phase, the duel was Porsche against Toyota. The race ended with a double victory for the 911 GT1 '98.

Porsche 919 Hybrid

Nico Hülkenberg

Earl Bamber

Nick Tandy

2015

Success for avant-garde hybrid
Porsche placed its faith in a unique energy recovery system and competed in the highest efficiency class. Two 919 Hybrid cars finished the race at the front.

Porsche 919 Hybrid

Romain Dumas

Neel Jani

Marc Lieb

2016

First place after continuous battle
There was nothing to separate Porsche and Toyota for 23 hours, 56 minutes and 35 seconds. The Japanese car then came to a stop. The Porsche 919 Hybrid won the race.

Porsche 919 Hybrid

Earl Bamber

Timo Berhnard

Brendon Hartley

2017

Not for weak nerves

On early Sunday morning, one of the two 919 Hybrid cars was forced to retire while in the lead. The other started an unbelievable chase through the field from 56th place. It ended with the 19th overall victory in Le Mans.

Number of employees

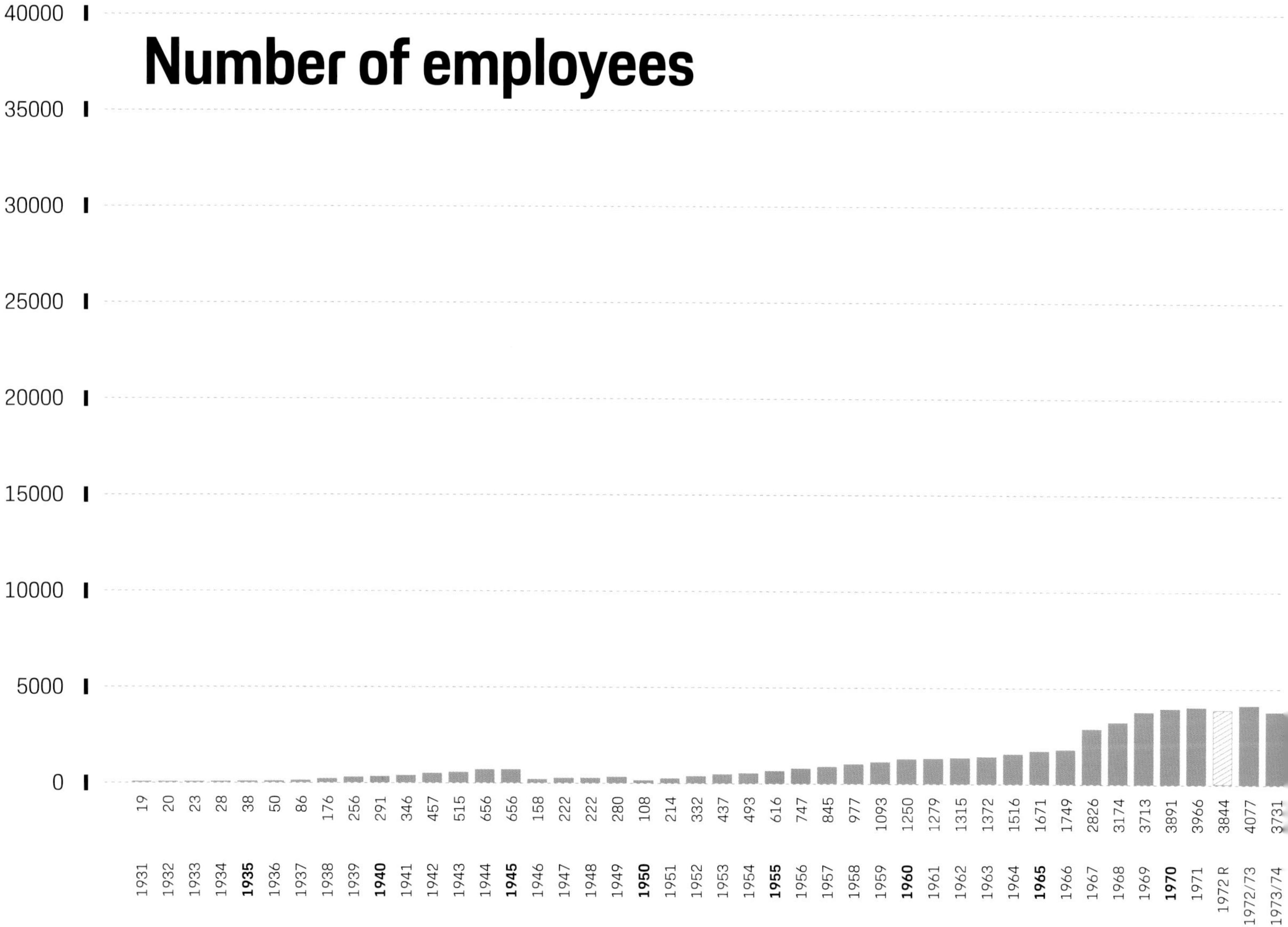

Short financial year: Due to the former Porsche KG being turned into a stock corporation on 1 August 1972, the financial years from then on went from 1 August to 31 July of the following year. Resulting from the creation of an integrated automotive group with Volkswagen, the Porsche AG financial year was changed back to calendar years as of 1 January 2011. For the periods of 1 January 1972 to 31 July 1972 and 1 August 2010 to 31 December 2010, a so-called short financial year was therefore established in each case.

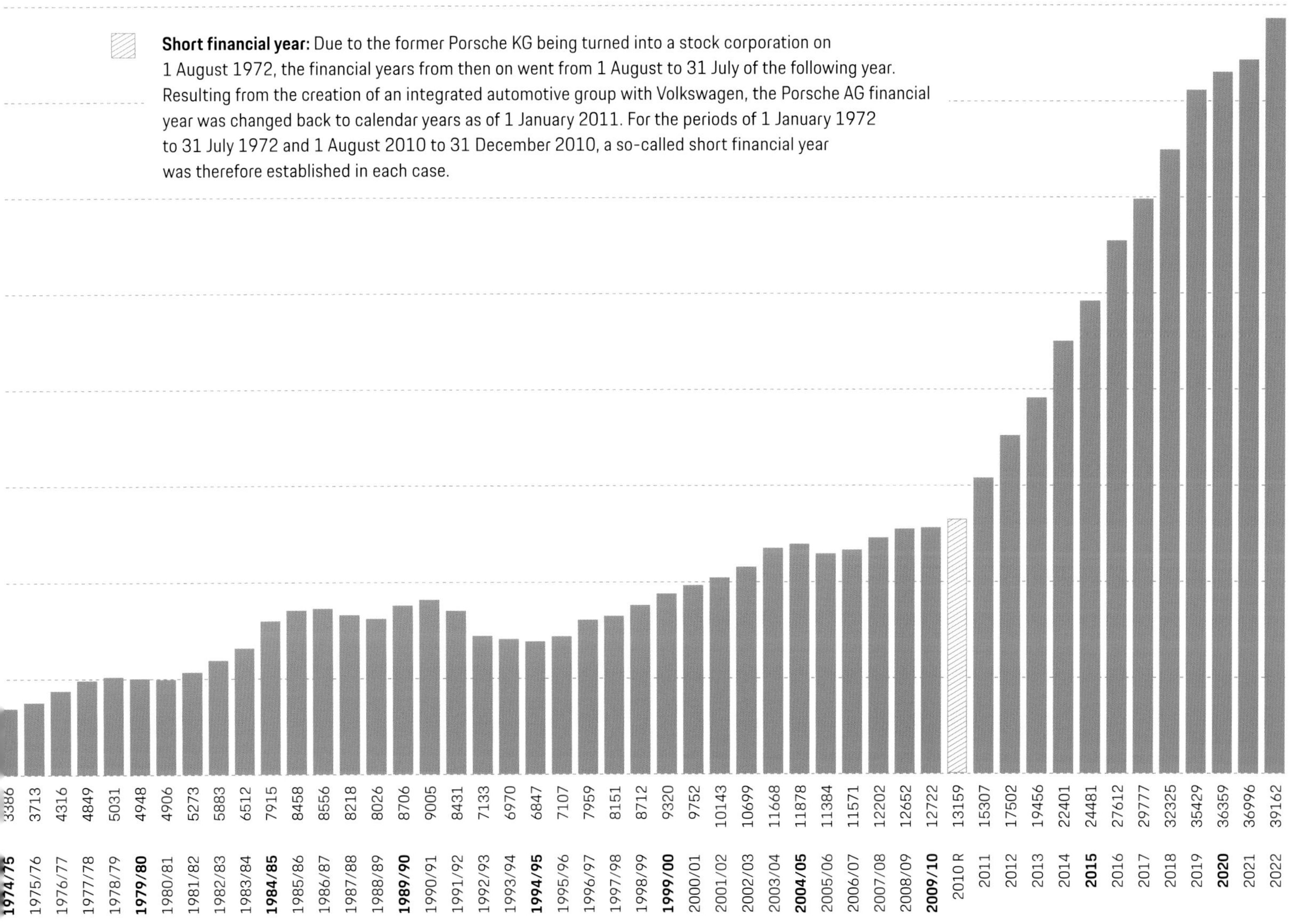

Business volume in millions

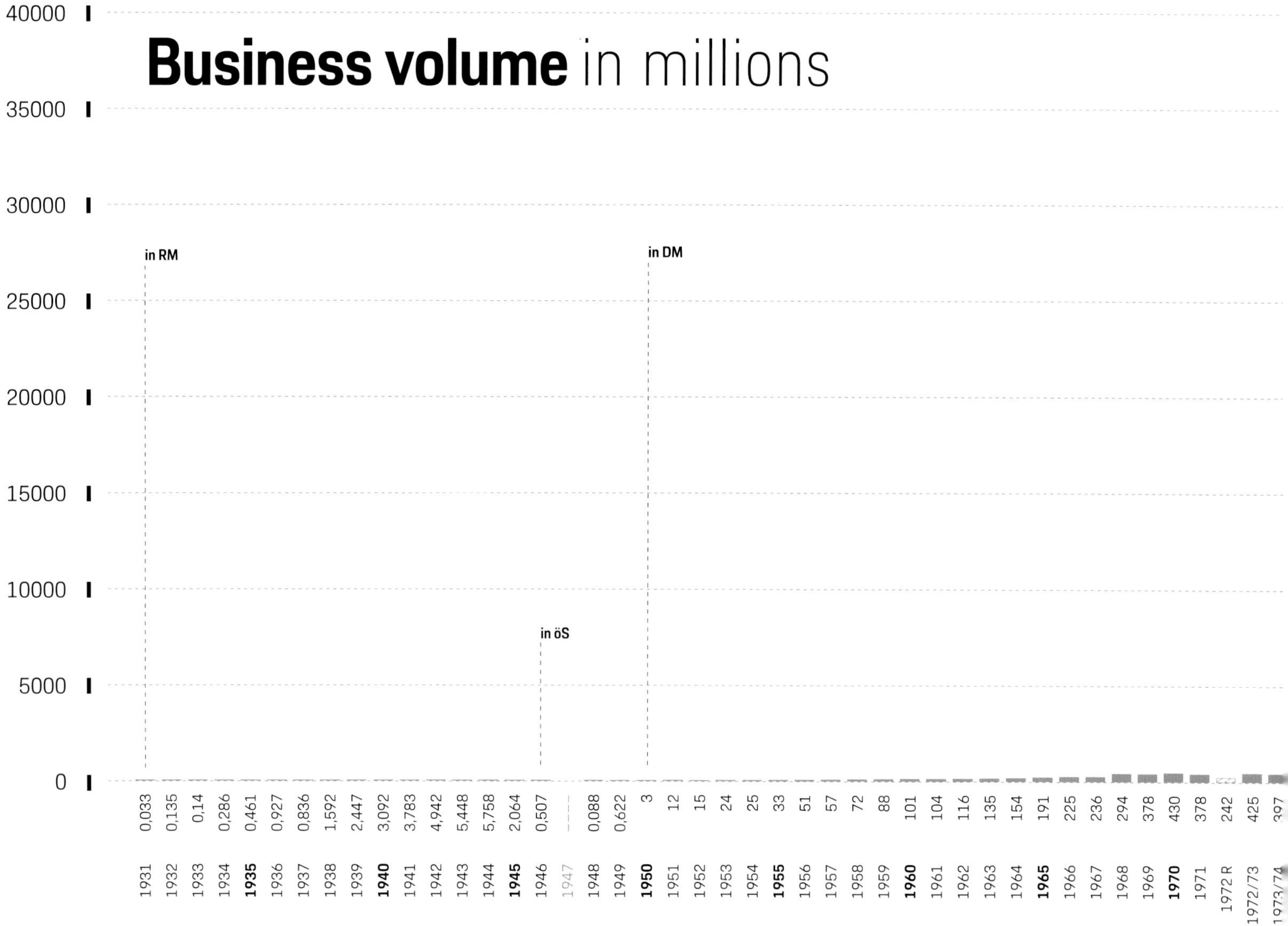

Short financial year: Due to the former Porsche KG being turned into a stock corporation on 1 August 1972, the financial years from then on went from 1 August to 31 July of the following year. Resulting from the creation of an integrated automotive group with Volkswagen, the Porsche AG financial year was changed back to calendar years as of 1 January 2011. For the periods of 1 January 1972 to 31 July 1972 and 1 August 2010 to 31 December 2010, a so-called short financial year was therefore established in each case.

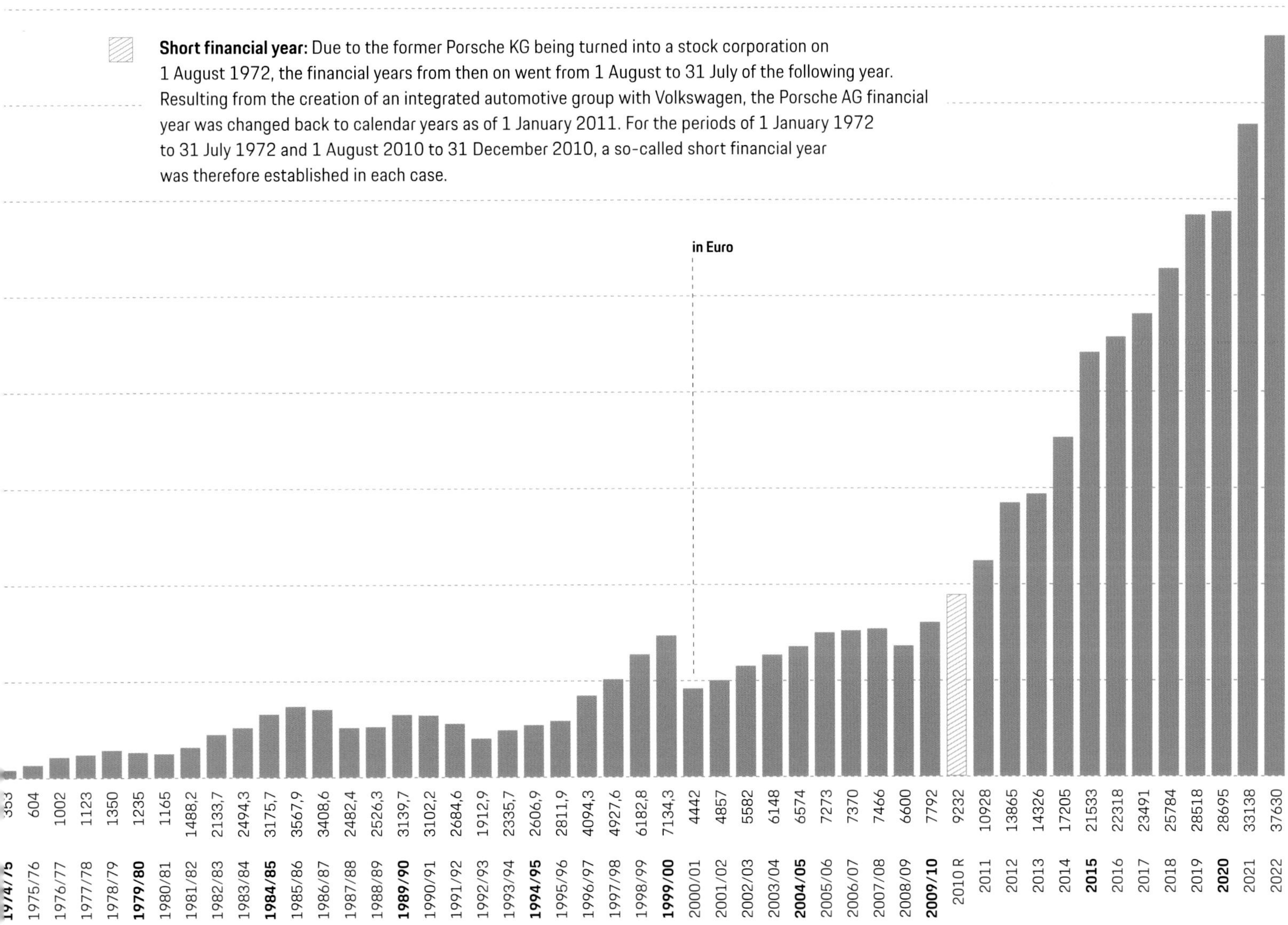

Production volume as of 1948

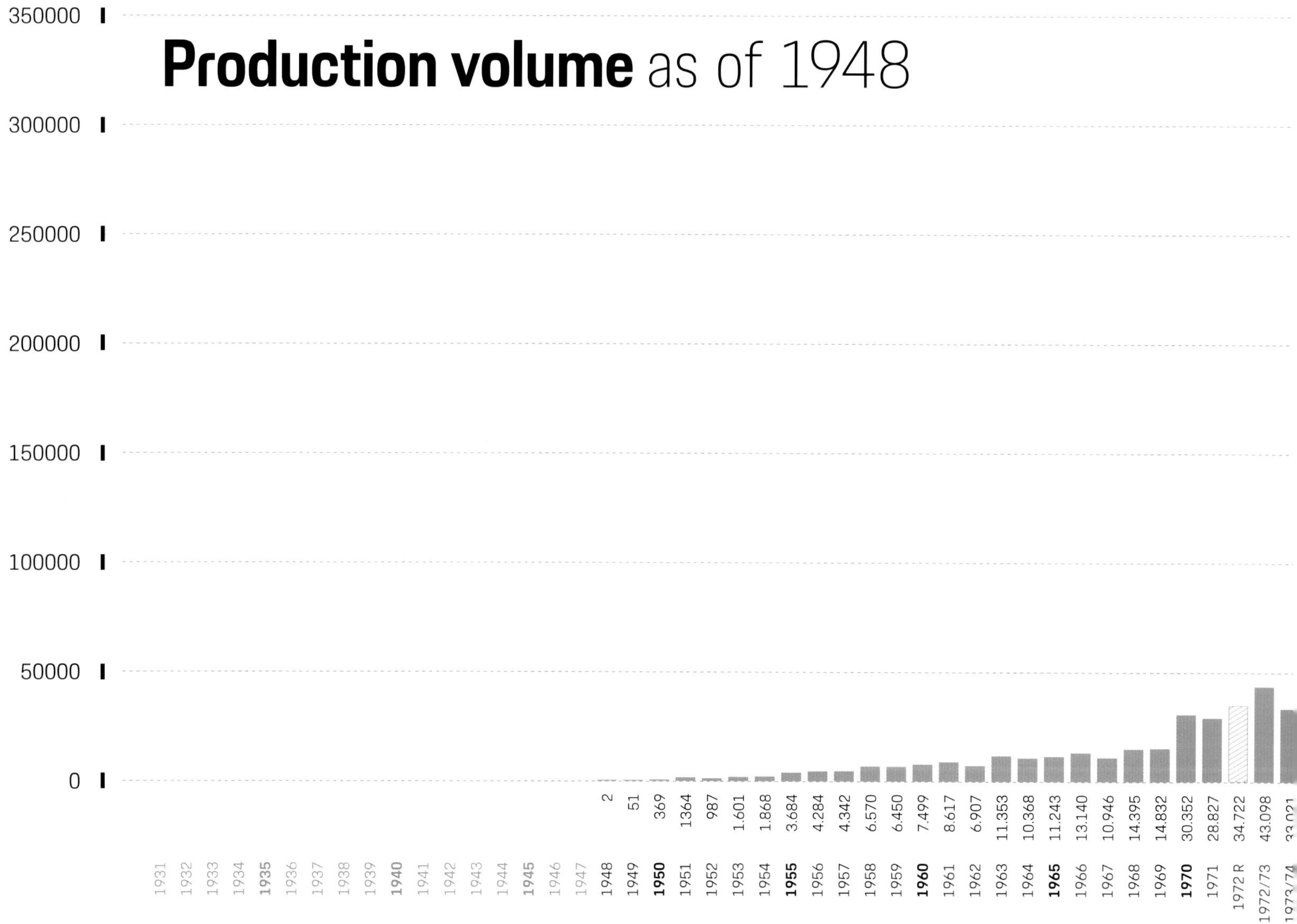

Short financial year: Due to the former Porsche KG being turned into a stock corporation on 1 August 1972, the financial years from then on went from 1 August to 31 July of the following year. Resulting from the creation of an integrated automotive group with Volkswagen, the Porsche AG financial year was changed back to calendar years as of 1 January 2011. For the periods of 1 January 1972 to 31 July 1972 and 1 August 2010 to 31 December 2010, a so-called short financial year was therefore established in each case.

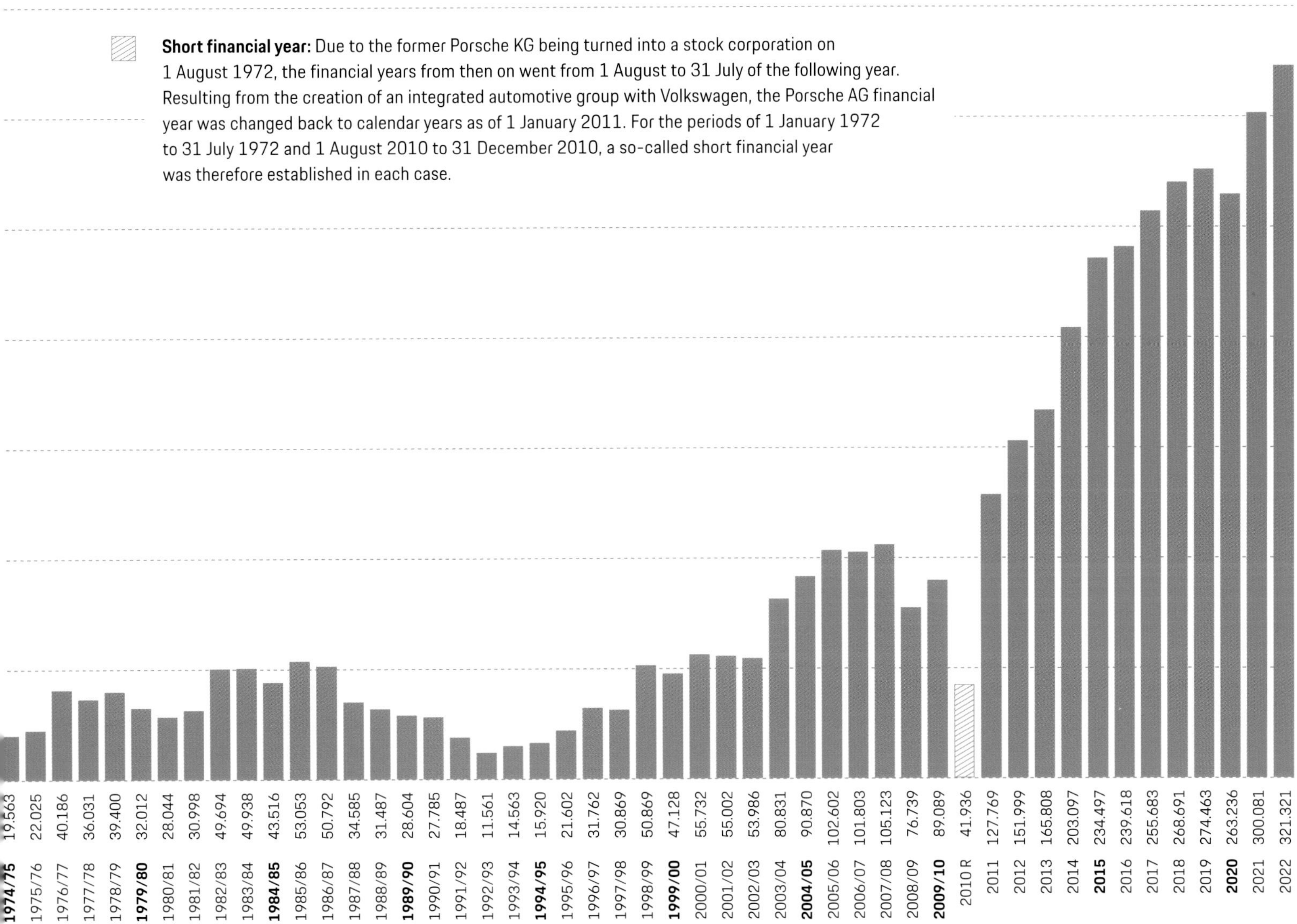

Porsche Experience Centre

Details of the fuel consumption and CO_2 emissions can be found on page 282 and 283.

INDEX Keywords

M

N

O

P

R

S

T

INDEX Constructions

Fuel consumption and CO_2 emissions

Status 04/2023

718 Boxster models

Fuel consumption combined (WLTP): 11,1–8,9 l/100 km;
CO_2-emissions combined (WLTP): 251–201 g/km

718 Boxster GTS 4.0

Fuel consumption combined (WLTP): 10,9–10,1 l/100 km;
CO_2-emissions combined (WLTP): 247–230 g/km

718 Boxster T

Fuel consumption combined (WLTP): 9,7–8,9 l/100 km;
CO_2-emissions combined (WLTP): 219–202 g/km

Boxster 25 Jahre

Fuel consumption combined (WLTP): 10,9–10,1 l/100 km;
CO_2-emissions combined (WLTP): 247–230 g/km

718 Cayman GTS 4.0

Fuel consumption combined (WLTP): 10,9–10,1 l/100 km;
CO_2-emissions combined (WLTP): 247–230 g/km

718 Cayman T

Fuel consumption combined (WLTP): 9,6–8,9 l/100 km;
CO_2-emissions combined (WLTP): 218–202 g/km

911 Carrera models

Fuel consumption combined (WLTP): 11,4–10,1 l/100 km;
CO_2-emissions combined (WLTP): 259–229 g/km

911 Carrera S Cabriolet

Fuel consumption combined (WLTP): 11,0–10,3 l/100 km;
CO_2-emissions combined (WLTP): 250–233 g/km

911 Carrera 4S Cabriolet

Fuel consumption combined (WLTP): 11,1–10,4 l/100 km;
CO_2-emissions combined (WLTP): 252–235 g/km

911 Dakar

Fuel consumption combined (WLTP): 11,3 l/100 km;
CO_2-emissions combined (WLTP): 256 g/km

911 GT3

Fuel consumption combined (WLTP): 13,0–12,9 l/100 km;
CO_2-emissions combined (WLTP): 294–293 g/km

911 GT3 RS

Fuel consumption combined (WLTP): 13,4 l/100 km;
CO_2-emissions combined (WLTP): 305 g/km

911 Sport Classic

Fuel consumption combined (WLTP): 12,6 l/100 km;
CO_2-emissions combined (WLTP): 285 g/km

911 Targa 4S Heritage Design Edition

Fuel consumption combined (WLTP): 11,1–10,4 l/100 km;
CO_2-emissions combined (WLTP): 253–236 g/km

911 Targa models
Fuel consumption combined (WLTP): 11,1–10,4 l/100 km;
CO_2-emissions combined (WLTP): 253–236 g/km

911 Turbo S
Fuel consumption combined (WLTP): 12,3–12,0 l/100 km;
CO_2-emissions combined (WLTP): 278–271 g/km

Cayenne
Fuel consumption combined (WLTP): 12,5–11,5 l/100 km;
CO_2-emissions combined (WLTP): 283–259 g/km

Cayenne Coupé models
Fuel consumption combined (WLTP): 14,1–11,6 l/100 km;
CO_2-emissions combined (WLTP): 283–263 g/km

Cayenne E-Hybrid
Fuel consumption weighted combined (WLTP): 3,7–3,1 l/100 km;
CO_2-emissions weighted combined (WLTP): 83–71 g/km;
Electrical consumption weighted combined (WLTP): 26,5–25,1 kWh/100 km;
Electric range (EAER): 41–44 km;
Electric range City (EAER City): 44–48 km

Cayenne S
Fuel consumption combined (WLTP): 12,9–11,8 l/100 km;
CO_2-emissions combined (WLTP): 292–268 g/km (WLTP)

Cayenne Turbo GT
Fuel consumption combined (WLTP): 14,1 l/100 km;
CO_2-emission combined (WLTP): 319 g/km

Macan T
Fuel consumption combined (WLTP): 10,7–10,1 l/100 km;
CO_2-emissions combined (WLTP): 242–229 g/km

Macan S
Fuel consumption combined (WLTP): 11,7–11,1 l/100 km;
CO_2-emissions combined (WLTP): 265–251 g/km

Taycan Cross Turismo models
Electrical consumption combined (WLTP): 24,8–21,2 kWh/100 km;
CO_2-emissions combined (WLTP): 0 g/km;
Electric range combined (WLTP): 415–490 km;
Electric range in town (WLTP): 517–616 km

Taycan Turbo models
Electrical consumption combined (WLTP): 28,5– 8,0 kWh/100 km;
CO_2-emissions combined (WLTP): 0 g/km;
Electric range combined (WLTP): 435–506 km;
Electric range in town (WLTP): 537–627 km

IIMPRINT AND LEGAL NOTICE

Dr. Ing. h.c. F. Porsche AG
Porscheplatz 1
70435 Stuttgart
Germany
www.porsche.com/museum

PUBLISHER
Dr. Ing. h.c. F. Porsche AG
Edition Porsche Museum 2023
MAP09076523

Motorbuch Verlag, Stuttgart
ISBN 9783613321618

BOOK DESIGN & LAYOUT
design hoch drei GmbH & Co. KG,
Mark Ch. Klein, Jonathan Schönberger

PRINTING
Printer Trento srl

PHOTO CREDITS
Photos: All photos by the Dr. Ing. h.c. F. Porsche AG Corporate Archives
except the following:
Adolf Rosenberger gGmbH (p. 12),
Automuseum PROTOTYP Hamburg (p. 40 right),
Porsche AG (p. 216 left, 219 top left, 221 right, 223-251, 254-255, 270-271),
Reutter Familienarchiv (p. 81 right).